Saline District Library
3 4604 91151 7737

P9-CRD-432

Also by David Edwards

The Lab: Creativity and Culture

Artscience: Creativity in the Post-Google Generation

CREATING THINGS THAT MATTER

CREATING THINGS THAT MATTER

THE ART & SCIENCE OF INNOVATIONS THAT LAST

DAVID EDWARDS

Henry Holt and Company New York

Henry Holt and Company
Publishers since 1866
175 Fifth Avenue
New York, New York 10010
www.henryholt.com

Distributed in Canada by Raincoast Book Distribution Limited

Library of Congress Cataloging-in-Publication Data

Names: Edwards, David A., 1961– author.
Title: Creating things that matter : the art & science of innovations that last / David Edwards.
Description: First edition. | New York : Henry Holt and Company, [2018]
Identifiers: LCCN 2018003349 | ISBN 9781250147189 (hardcover)
Subjects: LCSH: Creative ability in technology. | Technological innovations. | Creative ability. | Inventions.
Classification: LCC T49.5 .E39 2018 | DDC 600—dc23
LC record available at https://lccn.loc.gov/2018003349

ISBN: 9781250147189

First Edition 2018

Designed by Meryl Levavi

Printed in the United States of America

1 3 5 7 9 10 8 6 4 2

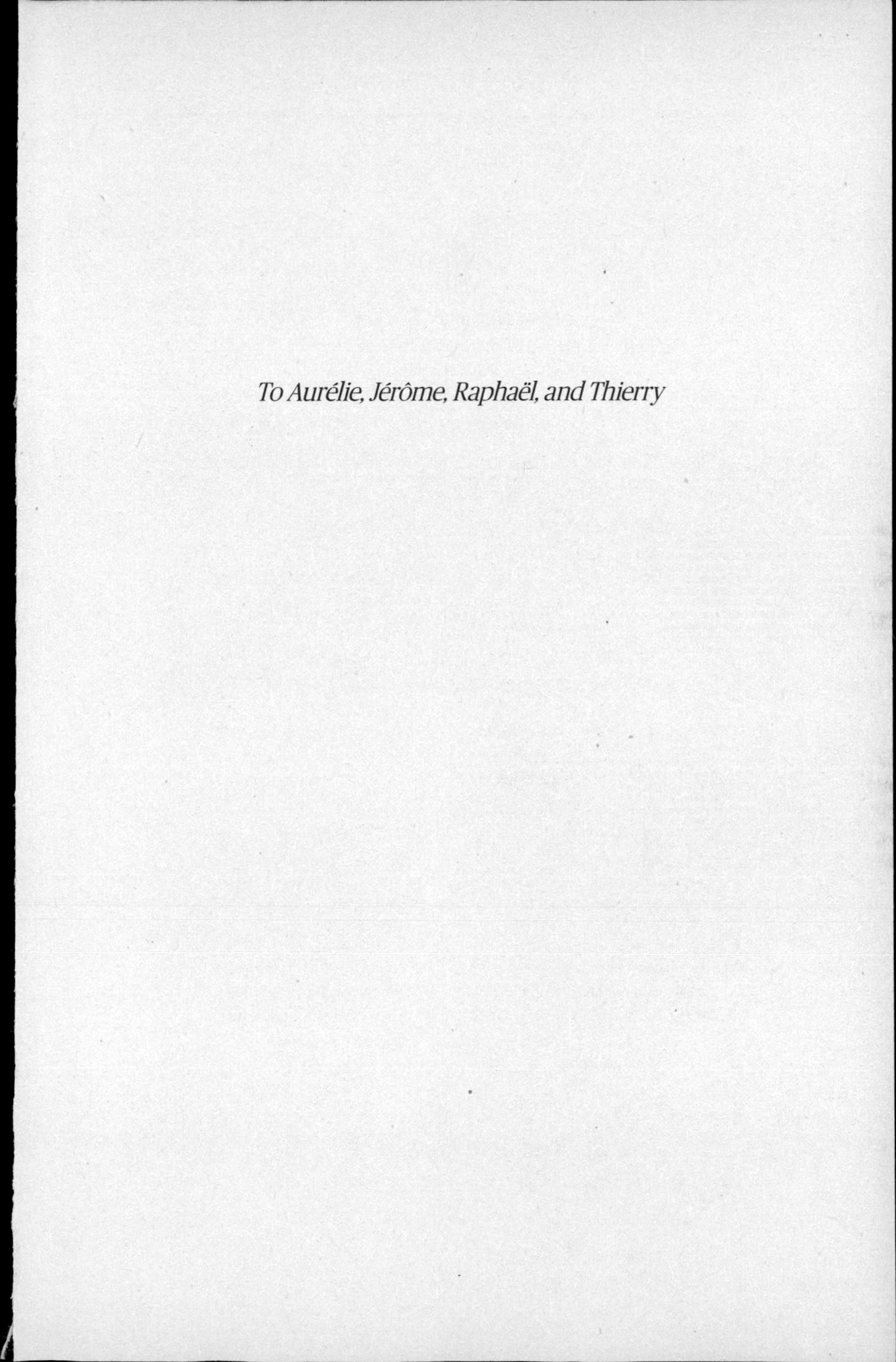

To Aurélie, Jérôme, Raphaël, and Thierry

Contents

PART III ◦ MAKING A FUTURE WE ALL WANT

Introduction

MOST THINGS WE CREATE WILL NOT MATTER. THIS BOOK is about creating things that do. It is about how we create things that bring enduring value to the planet.

We get the general idea of creating when we build our first sandcastle. Write our first short story in school. Tell our first convenient lie. Creating for our own purposes turns out to be an easy matter. It is when we create value for others over a long time that we evoke the idea of the sublime.

Creating very new things that durably matter is one of the most difficult things we may ever attempt. The process looks radically unlike the cutthroat stereotype of innovation success. Empathy counts more than selfishness, innocence more than experience, aesthetic intelligence more than engineering brilliance, and humility more than arrogance.

By aesthetic characteristics more commonly associated with artistic expression or creative play, artists, scientists, engineers, designers, entrepreneurs, and other creators pioneer frontiers, discover, and occasionally create things that change how we live and think.

Aesthetics matter—what seems *not* to matter is precisely what does.

Aesthetics are the ways of the mindfully engaged creator and the qualities of the remarkable created thing, like the Sistine Chapel or quantum mechanics. By aesthetics, art and science each advance frontiers, and in the exploration of the unknown, these famous opposites become the same.

Experiential aesthetics has roots in the social, technological, and cultural upheavals of the late nineteenth and early twentieth centuries. The Italian philosopher Benedetto Croce wrote in his *Aesthetic as Science of Expression and General Linguistic* that the common definition of aesthetics, as a set of principles defining the meaning of beauty, had lost meaning. It was nonsense to define beauty in painting, sculpture, architecture, dance, and other art forms by fixed rules amended by continual corrections. Aesthetics related more obviously to process than to outcome. According to Croce, it made art and science one. The American philosopher and educator John Dewey, in his *Art as Experience*, carried the argument further. Beautiful forms successfully express sensitive human experience. Anything might be beautiful. A novel, a dance, a bridge. The more universal the experience, the more associated with the very long arc of human existence, the more a finely achieved expression, or creation, seems to be art.

Creating Things That Matter is about how we can learn to create this way, what it means to each of us personally, and why it promotes the creating of a world we actually want to live in. It is about how we observe, dream, and act in a way exemplified by the practices of some of our most notable creators—a process that is at once new to our boardrooms and business schools and as old as humanity itself.

THE AESTHETICS OF CREATING

CHAPTER 1

Creating in a World That No Longer Exists

Dad was bigger than any father I knew. As long as we lived together, I was in awe of him. The electric trains I received for my birthday never entertained me as much as they did when he came into the basement to spend an afternoon with me building fantastic mountains our trains could climb and colorful villages they could meander through. The tomato I carried onto the porch with a shaker of pepper didn't taste nearly so good as it did when he walked out there with me and showed me how to sprinkle and bite, sprinkle and bite, with red juice running down our chins.

When my dad sat down to watch a football game, I instantly wanted to do the same, and when he roared with pleasure or fury, I roared too.

He loved to teach, and I suspected his students to be like a second family to him. It worried me a little. Whenever he invited his students into our house, I watched them carefully, as if some of my dad had rubbed off on them and not me, and by close observation, I might gather up these missing parts and be made whole. Once, when I was thirteen, he spoke over the phone to a student about a chemistry problem. The conversation went on and on, right there in the center of our kitchen where my mom sometimes made delicious chocolate chip cookies, and behind her back, my dad would snatch for us globs of dough. It was exasperating to me, this nonfamily chemistry talk. "Stop!" I yelled. I couldn't stand the idea that there was this piece of my dad I couldn't have.

He was my dad. He was mine and he was incomparable. Next to him I felt empowered, yet I also felt alone, as if I stood next to a brilliant warm sun that sometimes blinded me. When I left home, my dad and I saw each other less frequently. After college, grad school, and postgraduate studies, where I improbably moved from a community college to MIT, I started to travel, to do very new sorts of things, and discover worlds that my dad didn't know. I moved to Paris and opened my lab to the public, calling it Le Laboratoire, where I staged wild exhibitions that explored questions like the meaning of "now" with the South African artist William Kentridge, and flash culinary sensations with the French designer Philippe Starck.

One day in the midst of all this, I was invited into a radio studio for an interview. It was a rainy day, and I recall entering the studio on rue de Cléry with wet jeans and a distracted mind. Two technicians greeted me at the door, ushered me into a tiny room, and set me before a thick microphone. They placed a heavy headset over my ears. I heard a voice, deep and familiar. The NPR journalist, based in Atlanta, asked me if I'd like to sing. I laughed. What a question, I thought. No, I said, I had no singing talent at all, to which the man replied, Neither does my wife, and yet you just asked her to sing. His wife happened to be visiting Paris. She had experienced an exhibition at Le Laboratoire called *Vocal Vibrations* and had just phoned him from the Charles de Gaulle Airport. How amazing, I said. I hope she enjoyed it, but I still won't sing for you. We both laughed, and I opened up as I'm sure he supposed I would.

Midway along, the journalist asked me if I had a single person to thank most for my creativity. I thought about it. Yes, it was my dad.

He asked me what secret my dad had shared with me.

He had done one thing, I replied, and it had changed me forever. My dad got down on the floor and played with me. We made up games with toy soldiers and invented together. Lying there on the cold cement, we were collaborators, a kid and his dad, cocreators of things that mattered.

I choked up. Nothing like this had ever happened to me. Here I was before a complete stranger in the middle of an interview. It was crazy. This journalist had somehow gotten to the core of me. In the sound booth, I had gone back to creating with my father in the basement of our tiny suburban house, and it hurt me to have to leave.

Recollections like these can be powerful, all the more as childhood creation differs so fundamentally from how we create later on. As kids, we create a story, a toy house, or a personality, mindful and imaginative, with little clear idea of what our project will become. This adventurous approach—or "third way"—to creating often gets sidelined as soon as we enter school, where we learn to advance along one of the two standard paths of contemporary creation, the commercial and cultural paths, with their regulations and constraints.

In the commercial approach to creating, we try to discover and meet a consumer or general public need before anyone else does, so that we profit, and the world profits, at least for a while. Commercial creating is competitive, like a sport, with success supported by some combination of scientific discovery, technological mastery, commercial expertise, and reasonably free markets. The commercial model, by which we might create a new airplane or an original kind of shoe, requires significant resources and aims to pay these resources back in the span of a few years.

In the cultural approach to creating, we express personal experience and artistic inclination in forms—like new books, musical compositions, or choreographies—that may eventually change how people think and live. Cultural creation happens on a less specific time frame than commercial creation does. Cultural creating can cost almost anything, and while creators sometimes reap profits, their returns are generally less tangible, as in learning or some kind of humanitarian or cultural contribution. The goal of cultural creation tends to be personal, with reward deriving from the expression itself.

Commercial creating is like tossing a dart. We aim and throw, and either hit the target or miss it. Cultural creating is like hiking through an uncharted canyon. Riveting while meandering, it succeeds by leading us to places of surprise.

The dart-throwing method gets credit in our schools, corporations, and governments for working best. So rapid and effective have we been over the years with commercial innovation that it has become synonymous with what we think of today as creation. Even films, books, music, and storytelling games are largely now made using the same strategy by which we make a new cell phone or a car. They greet the public by way of commercial process, since near-term profitability matters more, in the bottom-line business of culture, than long-term transformative impact on how people will ultimately think and feel.

As to resource allocation, the commercial camp again beats out the cultural camp. Statistics reveal why. Thanks largely to the commercialization of scientific and technological advances, average human life expectancy has shot up since 1900 from around thirty to over seventy years of age, while the probability of infant death has fallen from 35 percent to less than 5 percent globally. Health inequalities between countries, ethnicities, and income levels have mostly fallen over this same time, and childhood malnutrition rates have gone from an estimated 35 percent in developing countries and 22 percent in developed countries in 1900 to less than 5 percent globally today. People all over the world now enjoy broad access to education, running water, waste treatment, information, entertainment, and more. On the whole, thanks largely to the commercial model of creation, we live better today than we ever have.

The things we have made to live comfortably have obviously mattered. Beyond the direct benefits they bring to us, they also bring indirect benefit by employing us in their economies of production and exchange.

Suddenly, however, our many inventions, from skyscrapers to polyester clothing, point to global crises. Our dilemmas range from disappearing jobs; growing scarcities of water, minerals, clean air, and other natural resources; mass animal extinctions; spreading plastic gyres in the ocean; rising sea levels; and widening health care and education inequalities. The sources of our problems go beyond the things we have made, how we have made them, and how we dispose of them when their utility is gone and lead to a host of consequences, some intended and many not. We have more children surviving and living longer, everywhere, than at any time in human history. Modern communication and transportation have blurred the lines of country and village, home and office, disrupting traditional ways of organizing ourselves into groups with common interests. A created thing, like an iPhone, can alter the lives of billions, while a tweet sent from an iPhone in Washington, DC, on a Sunday evening at seven o'clock, just before many people are sitting down to dinner, can generate news around the world and set nations on the path to war.

We have, of course, met tall challenges many times in human history. None of these obstacles may seem insurmountable given the creative solutions we've come up with in the past. But commercial innovation is clearly falling short of addressing today's challenges. For a few decades now, we have recognized the liability of poor air quality, which shortens lives; fossil fuel consumption, which heats up the planet; and sugar-derived food addictions, which aggravate the global obesity pandemic. Each of these problems has one or more scientific and technological solutions. Still, commercial creators (producers of cars, oil refineries, and food manufacturers, among others) have mostly failed to invest in the innovations that might promote a healthy human condition in the longest term, if simply because it is economically advantageous to avoid it in the shortest term. Commercial innovation cannot value what it cannot measure, and in the commercial model, the net present economic value of the distant future is negligibly small. Our cultural mode of creating, which invites us to feel the world around us and to create things that express what we feel, mostly flourishes outside the institutions we use today to change the human condition, with the consequence that its impact on how we think and live is not as great as it might otherwise be.

Besides this, changing the human condition today will be harder than it was in ages past. We live better, not worse, than we did a century ago. Creating things that entice well-fed people to eat differently for benefits they won't immediately feel will be more difficult than creating things that help people eat. Creating things that get people to manage their own health and wellness when they do not feel sick will be more difficult than creating things that heal wounds and kill disease. And creating things that help people connect to others and dialogue when they are already massively connected in modern social networks will be more difficult than creating things that connect people living in isolation.

In these and other examples of the challenges we face, we have less and less chance to point to personal benefit as a principal motivation to adopt new things. We need to point to an attractiveness or "beauty" that transcends personal benefit, disrupting markets whose basic needs are already addressed, as James Dyson did for home appliances (from the bagless vacuum cleaner to the bladeless fan), and Elon Musk did for the electric car (with Tesla).

Aesthetic creating, the "third way" to create, happens to be the most natural kind of creating there is. A sensitive encounter with the world leads to a new creation that expresses what we feel.

A child creates aesthetically when she twirls a pirouette before her older brother, doodles on a piece of paper during class, or writes a first love letter. She experiences the world with innocence and sensitivity, hoping, noticing, and feeling what older people often will not. She expresses her impressions with a created thing, the pirouette or the love poem. Seeing what she has created changes her in some way, as does watching how others respond to it. They may see in her doodle or in her love letter something she did not previously reveal, and they understand her better for it. Possibly they will be moved and for a little while think or feel differently as a consequence.

All of this can, of course, be said of Michelangelo's ceiling mural in the Sistine Chapel or Albert Einstein's theory of relativity, next to which the child's creation is a hint of possibility, like an aroma, transient and fragile.

Aesthetic creating may end up changing the world, but it starts in fragility, making it reliant on supportive culture.

It was out of gratitude for *this* that I lost my voice on the radio.

What We Mean When We Say a Thing is Beautiful

As I write these words, I'm sitting in Massachusetts, on a couch made in Italy, drinking an espresso from an Ecuadorian coffee bean, eating an orange from Florida, wearing a sweater made in India, and noticing a text message from my wife in Paris. I have three email inboxes with over twenty thousand messages, and I get pop-up messages on whatever screen I have open (phone, tablet, computer) from the *New York Times*, *Le Monde*, the *Wall Street Journal*, and other newsworthy sites. I can stop writing, turn off my brain, and mindlessly kill a whole day here, cared for and shaped by whims and forces from around the globe. Meanwhile, the world around me changes in leaps that no one seems able to predict—politically, socially, economically, and technologically. My conditions change due to various outcomes of human creativity by the day, and sometimes by the minute, even though I have not asked for this change. It just happens. It is as if there are a whole lot of people out there experimenting with me inside their lab.

Millions of experiments do, in fact, change the human condition in the course of a day. If we can step outside the blur of it all and not get caught up in the consumerism of our era, there can be excitement and discovery in the stream of newness around us today. We need to pay close attention to notice what is exactly particular to the change around us, beyond the tabloid "noise" of living in the twenty-first century. When we do, we tend to express original thoughts and feelings, by creating something—perhaps writing a letter, making a new blog entry, or inventing a new light bulb.

Aesthetic creating is in this sense not only a third way to creating tomorrow; it is a way of living and adapting to change.

Children are the heroes of aesthetic creating not least of all because their condition changes so rapidly as they grow up. They react to change from inside and outside, from the hormones that circulate within them to the parade of teachers, classmates, and friends who march past them. Children create personalities, scenes, phrases of speech, personal crises, games, and countless other things we may sweep away as part of the process of growing up. In truth, this creative process is as fundamental to their adaptation and autonomous emergence into the world as it is to our own.

When we create new things, from an original style of dress to a novel, we invest in these things an aesthetic quality that not only pleases us, but also hopefully pleases those around us. Pleasing others may take time, but if we stick with our creative process and make adjustments along the way, we might find that our pleasure grows. Paying attention to the feedback we receive from others about the things we create and changing our creations accordingly (it's not that we pander to opinion that is not ours; we actually listen to it, adopt it, and change our perspective) is a kind of generosity, as if we give to others some precious piece of ourselves.

How this works, and for many reasons doesn't work, gets to what we mean by aesthetics and what we have in mind when we call something beautiful.

The appearance of things has always mattered, if never for long in the same way. Symmetry, as in the Pyramids of Giza, was perfection in ancient architecture, while in the altered conditions of recent architecture, as in Frank Gehry's Guggenheim Museum in Bilbao, it has become unnecessary and even unnatural. Ornamentation was refinement in the baroque tradition, as in St. Peter's Square in the Vatican, while in the Bauhaus era, with its straight-backed chairs and minimalist houses, it became ostentation. To appear thin in a poor economy can be a sign of deprivation, while in a rich economy to appear thin can reflect a special attention to health and wellness.

What we call beautiful is subjective—it depends on who we are, what we have experienced, and generally on our particular place and time. As the study of the beautiful, aesthetics has therefore had a rocky history, one that is worth recounting briefly here.

While we have rarely had difficulty getting consensus in any particular society about what we perceive as beautiful, each time, place, and culture seems to point us to a peculiar perspective.

According to Plato, beauty of the transient kind (as in a work of architecture or the effective articulation of an argument) led to beauty of a universal and eternal kind. Beauty of the universal variety related to the good life and to ethics. Through constant self-improvement and by adopting the virtues of moderation, courage, wisdom, and justice, Plato claimed we could find self-completion in the things we create, from our children to our crafts and works of art. These things were beautiful because they guided us to the Good.

For Plato, aesthetics and experience were inseparable. Beauty was connected to experience.

The Pythagoreans and Hedonists, and later the Epicureans and Stoics, all sought self-fulfillment by an aesthetic experiential journey. The Pythagoreans and Stoics looked to the mind, hoping to understand the universe and control physical desire; the Hedonists and Epicureans looked to satisfying physical desire and reaping pleasure. They tried to live ethical lives while embracing beauty and from this obtain happiness, which Plato characterized as the possession of Beauty or the Form of the Good.

But what exactly was the Form of the Good? The question met blank stares. According to popular tradition, Plato once gave a public talk on Beauty as the Form of the Good, and the audience left in a fit of frustration.

According to rationalist philosophers, our perceptions fell into two categories: intuitions and deductions. The aesthetic expression of our intuitions was sometimes referred to as art, and the aesthetic expression of our deductions was called science. However, created things could also be a mixture of both, like the Parthenon, a work of art that is also a fabulous work of science and engineering. Michelangelo's Sistine Chapel fresco impressively expressed the forensic science of his day.

In the oldest aesthetic tradition, art and science hang from the same tree.

Western views of art, science, beauty, and human experience took a new tack with the scientific discoveries of Isaac Newton. Newton and his contemporaries showed how things moved in the universe according to prescribed unambiguous rules. Deductive perceptions suddenly had predictive power. With science we seemed able to foresee the future, or at least we could make an educated guess about what might happen. The trick was to know the conditions around which the future would be framed. Once you assumed to know these, everything seemed to follow, just as four followed two plus two. You could predict wind motion and design objects to fly. Deduce the principles of steam engines and eventually thermonuclear missiles. Understand and eventually create biological life. Obviously, all this was useful. For those who invested the resources to create new things, the sheer utility of science proved its superiority. Art, by comparison, seemed suddenly far less useful. As a result, it became natural to see science and art as polar opposites. One was practical, and the other wasn't.

Personal experience seemed increasingly less essential to how we created the future. Our feelings and dreams did not alter our ability, aided by the laws of physics, to predict what tomorrow would bring or to make things that would improve our lives.

Since Plato, philosophers had bundled together beauty, experience, and ethics into a larger field of study called axiology. In the post-Newton world, philosophers from Hume to Kant argued that beauty was distinct from our personal connection to it and that aesthetics was only useful as a means to objectively examine a work of art.

Western society meanwhile clipped along.

Political revolution swept across France, then Europe. From the start of the nineteenth century, the Industrial Revolution moved society's economic dependence from the farm to the factory. Populations went from an experience mostly in direct touch with Mother Earth to an experience mediated by machines such as the cotton gin and the steam engine. This multiplied productivity and ultimately concentrated wealth in corporations. Other innovations, like electrical lighting and basic sanitation, further increased the advantages of living in large cities, facilitating mass population migration, and flipping the old top-down methods of royal governance to more participatory bottom-up democracy.

Artistic expression in Western society now looked remarkably different than it used to. Impressionist painters from Manet to Pissarro expressed a human experience radically altered by train travel, urban conditions, and new technologies like photography, where time and space seemed to blur into each other. By then, official art circles had almost no tolerance for deviation from the rules, leading the renegade impressionists to set up their own exhibitions and start the avant-garde. After the turn of the twentieth century, cubism in art, and relativity and quantum physics in science, would carry these ideas of disrupted space and time even further.

Pioneers of art and science sensed the unity of frontier creation. According to the American pragmatist philosopher John Dewey, humanity lived experimentally, and the sensitive expression of experience was itself art. In his book *Art as Experience* Dewey posited that a common aesthetic experience underpinned art and science as complementary ways of perceiving and expressing the condition of being alive. Marcel Duchamp's famous 1917 work *Fountain*, which was essentially a repurposed urinal, could now be said to be beautiful, as indeed could the Empire State Building. Pioneering artists and scientists, from the composer Igor Stravinsky to the physicist Erwin Schrödinger, embraced the new aesthetics, while within schools, universities, corporations, and governments, the division between art and science remained.

The scientist and novelist, C. P. Snow, attacked this division in his famous 1959 Cambridge lecture, "The Two Cultures," arguing that the great problems facing the postwar twentieth-century world would never be tackled without the institutional reemergence of a single culture. Snow's lecture and book of the same title became an instant classic. Snow argued that not only did modern scientists stand on more mobile terrain than they'd long thought, but also the complexity of contemporary challenges, ranging from the future of energy to multiculturalism and resource inequality, overwhelmed the ability of anyone to assume an understanding of future conditions. Snow argued that these conditions were about to change in fully unpredictable ways and that scientists would now be lost without help from humanists, and humanists would be lost without help from scientists.

We were all at the frontier together. Without a single culture that permitted intuition and deduction to mix as it naturally does with experimentation, civilization had no chance.

Snow's argument made sense, although it hardly changed how society worked. The efficiency of production and consumption continued to propel our corporations, our schools, and our governments. If you lived and worked at the frontiers of human experience, you might embrace the new experiential aesthetics, and if you did not, you likely embraced the old.

The fantastic successes of pioneering science throughout the twentieth century made those who turned to science to win wars and drive the economy feel ever more confident in the certainty of their way of doing things. A cascade of pioneering research continued; no truth was sacred in particle physics, biology, nuclear chemistry, neuroscience, artificial intelligence, and more. The flow of scientific articles and discoveries made science process and outcome seem increasingly identical. This happened in frontier art too. The works of artists such as Jackson Pollock, Andy Warhol, and Olafur Eliasson became valued almost as much for how the artists created as for what they created. Art studios grew in size, and production rate grew as it did in science labs. Pioneering contemporary art, like pioneering contemporary science, explored ambiguous and evolving human realities.

All this seemed lost on the world of large organizations.

There was the truth of the frontier. And there was the other truth. Detached from frontier experience, art promoted the wildly commercial, while science, separated from aesthetics, entered the service of short-term economic and political interests, promoting the sustainability crises of today.

C. P. Snow had been right at least in this respect. We *were* all at the frontier. Beauty and human experience were, and are, intimately connected.

Our Renaissance

In the midst of our multiple crises of sustainability, stubborn as we are in holding on to views of human productivity whose benefits are ever shorter term, we happen to be on the verge of an unusual grassroots renaissance. There are a few reasons to believe this, as I explore throughout this book.

First, humanity has over these last years dramatically increased its public expressiveness. Creative expression links billions of people today, from the developers of virtual content to the makers of all kinds of material objects. We may be tweeters, bloggers, or composers of ballets (virtual things), but we may also be the makers of robots, pottery, food, wood works, and 3-D printed objects (material things). While much of this Grassroots Creator Movement involves intermittent creative activity and hardly belongs to a creative life commitment, it often tends to follow neither strictly commercial nor cultural paradigms, but a more intuitive or third way of creating grounded in ambiguous, exploratory, personal experience and ending in a beauty the creator perceives in her own private way.

Second, grassroots creators are increasingly being supported by "activators" of every social, economic, and geographical position. These activators, who range from supportive friends and family to major philanthropists, are acting in the way of patrons of the Italian Renaissance, who supported creators of the era like Leonardo da Vinci and Michelangelo. Unlike Italian Renaissance patrons, activators today more typically act with far greater hands-on presence. These activators are distributing unprecedented time and personal resources toward grassroots creative efforts that address long-range sustainability challenges of the kinds I next describe.

Third, specialized environments or "culture labs" are being set up for grassroots creating that share deep commonalities with those that appear in my stories of towering creators today. These culture labs help lifelong creators learn and retain the "intuitive and deductive," "imaginative and analytical," "art and science" capacities of the da Vincis of our era. By pursuing the third way of creating, deeply committed artists, scientists, chefs, entrepreneurs, designers, and others hone emotive and cognitive states that promote frontier survival. These states are precisely those that grassroots creator communities are beginning to teach in short-term maker environments around the world.

Will any of this matter to the collective human future?

To start to answer this question, I conclude this first chapter with a quick look at how and why fast creation, the prominent innovation model of today, came to shape how we eat, manage health care, and communicate—

three critical realms of human experience—and why the slower third way of creating will be needed to reshape these same experiential realms in ways that we can sustain.

Creating How We Eat

Most of the food we eat has been created to some degree by human intervention. Even the common tomato is a product of European and American crossbreeding (tomatoes used to be small, hollow, watery wild fruit). It is the nature of this intervention, with the benefits and consequences of the human touch, that now makes the food system at once short-lived and hard to change.

Two centuries ago, the trillion-dollar food industry would have been inconceivable. Families, villages, and cities tended to feast and starve together. People shared more. The rich ate well if they could, while the poor ate whatever came in relative abundance. Bread, often made with more sawdust and cow dung than grain, frequently kept the poor alive in Europe, which is why, with nineteenth-century urbanization and population growth, creative minds turned to making bread accessible and abundant.

From the invention of sliced bread in 1912, by Otto Rohwedder of Iowa, to the spread of Wonder Bread across the United States in the 1930s, a kind of miracle happened in American kitchens. Suddenly, it didn't matter where you lived, whether it was in the center of Chicago or in remote Wyoming, fresh abundant natural bread made it into your home. To make this happen, people invented chemical additives, which belonged to the family of molecules called oxidants and shortened the fermentation process. Others invented cellophane, the first food-grade polymer, which commercial producers adopted to keep Wonder Bread fresh from the time it was made to the time it showed up inside homes, wherever they were. Later on, more chemicals were invented, and better polymers produced. All this improved the consumer experience of bread and benefited the companies that made it. Similar innovations transformed other foods.

From food additives and plastics to herbicides and pesticides, innovators created things that made it possible, through commercialization, for much of the human population to eat what was essentially good tasting, reasonably nutritious, and abundantly plentiful food. Humanity benefited, as malnutrition rates fell and life expectancy grew.

What commercial innovators of the modern food system did not imagine, in their drive to solve our short-term challenges, was that the longer-term consequences of its advances would weigh so heavily on human health and the environment that we would need to abandon, modify, or replace their inventions even as they continued to serve people around the world.

In 1988 scientists first reported a large mass of tiny pieces of plastic floating in a gyre in the middle of the Pacific Ocean. By 2014, the Great Pacific Garbage Patch had grown to over 1,700 miles in length from edge to edge, weighed around 270,000 tons, and contained roughly 5 trillion pieces of plastic. This made it just one (albeit the largest) of several gyres by then known to be floating in the earth's oceans. Plastic waste, about half of which comes from food packaging, according to the EPA, can take decades and even centuries to degrade and dissolve into natural habitats. In other words, it enters the ocean's ecosystem much faster than it biodegrades. Ingestion of plastic, among other factors like overfishing, has decimated marine animal life over the last thirty years. The disappearance of life below the surface of the ocean reflects what is happening above it. Due to pesticides, insects are disappearing in alarming numbers. In the German state of Westphalia, entomologists have been setting up tents as traps and measuring insect biomass since 1989. Between 1989 and 2014, average biomass collected from May to October fell from 1.6 kilograms (3.5 pounds) to 300 grams (10.6 ounces). In a 2014 report in *Science*, researchers concluded that the earth had entered its sixth era of mass extinction with precipitous declines in populations of invertebrates and vertebrates alike. Beyond this, arable land is nearing its production capacity. Most of the future arable land needed to feed the projected 9 billion human population in 2050 exists in Latin America and sub-Saharan Africa, where population growth, deforestation, pollution, urbanization, and policy decisions all threaten what land we have left.

In short, we have created a nearly miraculous food system, on which most of us are dependent, and now we need to completely rethink it. The massive industry of food production, distribution, retail, and marketing that brings us our contemporary food system generates problems that many creators of our food system never imagined or, if they did, could not afford to think much about.

Suddenly, however, they can. Our obvious food sustainability challenge has garnered attention and resources. Many commercial innovators today are acting to help us change how we eat. Impossible Foods, a startup in San Francisco, uses chemistry and engineering to make vegetable-based meat that tastes, smells, and feels like meat. Indigo, a startup in Cambridge, engineers bacteria on crops to increase farm productivity. Original Unpackaged, a Berlin-based retail startup, sells food without packaging. Commercial insect farms have appeared around the world to grow edible protein of apparently inexhaustible supply for agricultural purposes.

These and other new products and services will surely matter to the future of how we eat, but our planetary challenge actually goes beyond technologies for cleaning up plastic from the oceans, growing insects, or coming up with new ways of growing and producing nutritious food. Our tallest challenge may now be to make everyone, and not just the affluent and educated, actually *want* to eat differently, everywhere.

We rely on the food system as it is. To change it now means we all need to move away from a system that works. This is a very different kind of challenge than we faced a century or two ago. Eating sliced bread packaged in a plastic bag was a benefit back when the alternative was stale bread or no bread at all. The solution to our sustainability problem may not be as appealing.

Yes, we need to invent the future of food, but for this

future of food to matter to us all, it will need to be beautiful.

Creating How We Stay Healthy

Two centuries ago, bloodletting, emptying out of the stomach, mercury, and various mysterious potions frequently met sickness and infection with more violence to the sufferer than the illness itself. By the mid-nineteenth century, some began to conjecture that washing hands helped prevent infection, although doctors didn't know why, and doubted it, until late in the century when Louis Pasteur, who happened to be a talented artist, finally demonstrated the undeniable existence of germs. Soon after came the first vaccines and antibody therapies, the working basis of aspirin, and powerful pain relievers like morphine, heroin, and codeine, which were all created from the opium poppy leaf.

When the twentieth century arrived, infectious diseases, including flu, tuberculosis, and gastrointestinal infections, were still the biggest killers of Americans. A child born in 1900 could expect to live until forty-seven years of age. That age went up, and the menace of infectious diseases went down, as improved understanding and methods of chemistry and biology made the empirical art of medicine look increasingly like a science. Consequently, by the end of the twentieth century, mortality rates had fallen over 50 percent, and the major killers of Americans had flipped to cancer and cardiovascular disease.

By the dawn of the twenty-first century, we had created techniques for manufacturing therapeutic proteins like insulin and drug delivery systems like needles, pumps, pills, inhaled drugs, and transdermal patches. We had created kidney dialysis machines and artificial hearts and highly functional prosthetic limbs. We had promising cancer therapies and sophisticated diagnostic capacities and hospitals that ranged from the primitive to the space-age. Most of this innovation was spurred by the commercial innovation model.

Meanwhile, the massive global investment in health care that propelled the creating of things with fantastic benefits also led to skyrocketing prices that soared well beyond what the average private individual can afford to pay to lead a healthy life.

In 1960 the average American had less than two hundred dollars in annual health care expenses compared to over nine thousand dollars per average American in 2014, an eightfold increase in inflation-adjusted currency. Over this same time frame, the percent of the US federal government budget devoted to health care expenses was on the rise. By 2016, health care expenditures by the US government had grown to an incredible $3.3 trillion.

Through commercial innovation and translation of science and technology advances, the modern health care system has saved many millions of lives. A century ago, heart failure, tuberculosis, cancer, and even a severe case of the common flu inevitably led to death. Today, struck by any of these conditions, we have a chance to stay alive, and possibly we will go on to enjoy excellent long lives. But with escalating costs, this chance is highly contingent on our income, our ethnicity, and where we live. Either we now accept a highly unfair distribution of health care access and continue advancing health care science and delivery within the boundaries of the systems we have today, or we create new ways of managing health and alter a giant health care system on which many are now reliant.

How we choose the path toward equitable health care is not so obvious. Scientific and technological advances, from gene editing to digital medicine, are promising fundamental changes to health care that might radically diminish the burden of disease and clinical care. But getting these to people everywhere necessitates a popular change in how we think about health care itself. Somehow, to make modern health care accessible to all, we need to adapt it to a broadly accessible way of living, making it independent of a doctor's prescription, so that it serves us in the full blush of good health. One of today's priorities is moving health care out of the clinic and into our lives through inventions, ranging from digital wearables to engineered foods, which we voluntarily integrate into our daily routines. We need to *want* these things, more enthusiastically than we wish to go to the hospital for our annual checkup.

Creating How We Communicate

Through most of human history information came to us from extremely local places. We sensed the immediate world around us: smelling, seeing, tasting, touching, and hearing it. We learned from what we sensed and teamed up with others to try to meet our daily needs, sensitive to the environment that helped support them. Information might come to us from other places and times, as oral stories, books, songs, theatrical plays, operas, and other forms of culture, but mostly we communicated in tiny communities of exchange. This meant that as a habitant of the northern coast of California, we could not know much of how people in Philadelphia thought or what they intended to do. We could live in Paris and not know much about what it meant to live in Vienna. We gleaned some things through works of culture and the mouths of travelers, though the larger world remained mostly a matter of the imagination. Our isolation had its benefits and its faults. Among the faults, if someone in Philadelphia planned to colonize our stretch of California, or if someone in Vienna wished to declare war on those of us living in Paris, we couldn't know. Not knowing what was happening in real time limited what we could know about our future and the control we could have over it. On the other hand, we paid closer attention to those around us, shared more common beliefs and values, and more coherently expressed these in works of culture. This benefited social cohesion, helping us survive.

Things couldn't be more different today. What started with creators tinkering across the nineteenth century to help us keep in touch on special occasions began to spill into our living rooms in 1920, when the first radio news, entertainment, and sports broadcasts appeared, and everything we might have learned by heading to the town square now came straight to us. The global communications market, which brings us GPS, the Internet, and global cell phone coverage, is somewhere between the size of the total food and health care markets. So useful is all of this that it is hard to imagine living outside it anymore.

By 2015, the average American spent over fifteen hours per day tapped into media of some sort. This extraordinary number, surely inflated by the propensity of many of us to plug into multiple media outlets, suggests that, aside from sleep and a little time to eat, speak to family and friends, and do chores, we are immersed in or at least influenced by a sensorial experience that is largely disconnected from our immediate physical environment. We listen to it, watch it, play with it, speak to it, dream with it, reel from it, and build our future with it not only on our mind, but almost as if it *were* our mind. Whatever sensorial stimulation we used to seek in a stroll down Main Street, coming home for an evening or passing a quiet night in our living room is now available, or at least its shadow is available, in a virtual world of mixed human and artificial intelligence. We are sharing our lives with it, in a way downloading our experience and uploading another, a cognitive and emotive experience framed by a digital screen and two sensorial dimensions. We pass our virtual existences mostly in a digital light and sound environment having evolved on the planet with all of our senses. What this means to our health, to our way of working and being, or to our ways of socializing, organizing ourselves in useful ways, or generally to our everyday behavior is not entirely clear, although rising rates of anxiety, depression, and ADHD, as well as social, political, and religious instabilities, are probably linked. Modern communications have altered the way people vote and the way politicians behave, and, of course, how we learn.

How we adapt to this new connected world is not only a question of socially and politically adjusting to what we have created, but of creating things that will help us adjust and benefit in durable ways.

What is true about the future of food and health care is true about the future of communications as well. Whatever we create to optimize new communication and make sustainable a massively connected global human society will not principally be meeting needs that went previously unmet but will be addressing needs in new ways. If they are to be massively adopted, they will need to attract us by their beauty.

What to Do

Our challenge is so gigantic that our condition can seem fated, as if we need to just learn to deal with it. The notion that global systems are imperfect can appear irrelevant to our own personal lives. It's easy to mindlessly drift through our days as if we can do nothing about them.

Of course, our biology has not evolved to promote mindlessness. Our brains help us survive and thrive by taking whatever touches us extremely personally. Biology encourages us to care about the world in which we live, and when we don't, when we assume we have no control over or input into our environment, we lose the millennia-formed equilibrium between our bodies and our environment. We overthink, with the result that emotion and cognition tug in opposite directions, and we make bad judgments. We lean too heavily on learned knowledge, and as a consequence, we split up the tasks of learning and doing. We fail then to learn as we might or do as we should. Our brains, contemporary neuroscience shows, naturally function better when their emotive and cognitive regions are engaged in active dialogue and by synthesizing the acts of learning and doing. Our biology inclines us to be mindful.

The answer to the question *what to do* in a challenged world is finally intuitive. The future is never deduced. In the world that actually exists, tomorrow gets pioneered by the intuitive lives of artists, scientists, and little kids playing on basement floors.

Creating Things That Matter is about leading this sort of life.

CHAPTER 2

Creating in a World That Does Exist

I CAME UPON THE THIRD WAY OF CREATING AS A TWENTY-three-year-old doctoral student in Chicago. MIT professor Howard Brenner had come to my university to speak on his favorite subject, fluid mechanics. Fluids, such as vinaigrette and olive oil, pour out of bottles, mix up in crazy ways, and flow over fresh garden leaves by laws of nature that took centuries for mathematicians, physicists, and engineers to figure out. Knowing those laws today makes it possible to design important things, like an artificial heart, and actually predict how they will work. Few scientists can say too much about fluid mechanics. Howard was able to say a lot during his visit, and I did my best to follow. If the sophistication of Howard's applied mathematics lost me in its detail, the elegance of his handwritten slides, the perfection of his symbolic language, and the naturalness, almost homeliness, of his manner fascinated me. I had the sense of observing a work of beauty (while before this moment mathematics seemed to me mostly functional, a means to an end). Howard was as comfortable in his rigorous incontrovertible mathematical world as I had once been in the make-believe world of my basement.

Years later, after I'd moved to Boston to study with Howard and we'd published our first book together on the movements of fluid surfaces, like those you see between the vinaigrette and the little droplets of olive oil, I began to wonder about the professional life I led. I loved what I did and might have been happy writing papers and books on applied mathematics for the rest of my life, but I had difficulty explaining my work to friends and family. This had never much bothered me. As I grew older, however, I became sensitive to the reality that my passions had become a life choice. Questions from friends and family turned from what I did to what I might eventually do. My difficulty in being able to express the joy of my work to anyone who could not understand my equations—meaning everyone I had grown up with—became a personal worry. I might have made more money, sought out more obvious adventure, and generally optimized my circumstances better had I followed a path like those of my old friends, who had gone to work as corporate engineers or salespeople and generally as professionals who delivered value to others in ways far more easily explained.

One day, I asked Howard how I should describe the nature of our work to others. Yes, we'd published a book. But just pointing to this book did not tell them of the daily adventure we shared. To others, it looked like the homework in school they had most loathed. What was this magical process we shared?

Aesthetics—he said. *We do aesthetics.*

I had assumed aesthetics related to the study of beauty. To the process and perception of art, not science. Aesthetics seemed to be what part of my brain did, the sensitive and vulnerable part that woke me up each morning to write a story and made my life bigger than it was. *That* was aesthetics. It tied back to my childhood play, to fictional games in the solitude of my basement.

The notion that my scientific work was guided by aesthetics, by deep and sustained engagement of the senses with a very personal world, surprised me. Howard was saying that the importance of *what* we did fundamentally related to the *way* we did it.

We made formulas that excited us. They involved the arcane language of polyadic math. Lowercase boldface letters stood for vectors, quantities that included both a value and a direction. You could express in a vector the idea of a car traveling fifty miles an hour headed east. Uppercase boldface letters stood for dyadic tensors, quantities that included a value and multiple directions. An example of a dyadic tensor was the idea of stress, how when you put your hands around a basketball and tossed it, you threw it in a direction and made it spin. With our polyadic formulas, we could express the laws of motion of the physical world in sentences that combined variations of lowercase and uppercase letters, with operations (for example, multiplication, subtraction, or inversion) between each letter showing up as dots, colons, and crosses. The appearance of the letters—dark-light-light-dark—with the other symbols showed you, when you understood this remote aesthetic language, whether the equation was true. It was legitimately beautiful . . . to the few who could understand it.

This valuing of my work beyond the usefulness of its outcome so contradicted what I'd been taught to see as the purpose and meaning of a scientific life that I kept Howard's observation to myself. He had answered my question, and it relieved me personally. But I didn't understand his answer well enough to wield it with my friends and family when they asked the same question of me.

Working with Howard made me comfortable about being *me*: a dreamer. But I still needed to be convinced that this *me*—this way of being shaped by a way of creating—mattered in the world.

What Matters

In the spring of 1993, my second book had appeared, and I had accepted a verbal offer to join the faculty of Chemical Engineering at the University of California at Santa Barbara. That same spring, the state of California ran out of money, the university froze hiring, and I lost the offer. I had no job. Walking the halls of Building 66, I looked past open doors at professors and imagined thousands of others just like them in and around Boston. Smart and fully employed people. Next to them, I had become an applicant. Each year they reflected on their faculty needs. Every few years they might invite applicants. Among those now would be me. The expression of me had turned into my CV. Few would read my books. Few could. I should have considered that before! My books were so specialized they had no practical audience. They now seemed to me remote and unwanted, like me.

The work I had perfected with Brenner had mattered to me. Evidently, it didn't matter so much to others.

Of course, countless things matter. Rainfall on a summer afternoon. Headlights in the dead of night. Poetry.

None of these things matter to all of us at all times. They each matter in some sense that we generally take for granted. Should the rain stop, headlights cease to function, poetry go away, we would be ruined. We need these things. Not always, but frequently enough that we support their endurance. We buy headlights. Write and read poetry. Fight for an ecosystem that will continue to deliver rainfall on a summer afternoon. They endure because we do better by them. Through these things, we eat, remember, share, create, stay healthy, understand, and care to live on the planet.

Creating what matters and has not existed before starts with what matters to us. We don't create, or we do so very poorly, by coercion. We create because we *feel* like it.

Teresa Amabile, a renowned creativity researcher at Harvard University, discovered early in her career a secret to diminishing creative output. She studied the motivations of writers. Writers, like any other creators, may follow the first, second, or third ways of creating: they might write for commercial benefit (to make money), for cultural benefit (to successfully publish a book), or for the benefit of passionate curiosity (to explore, like a pioneer). They may even follow all three ways, as many writers do, starting to write with passionate curiosity, crafting a book, which they hope to publish, and finally, having signed a publishing contract and while working on the final draft of a book, sitting down with marketers in the interests of having a commercial success. Amabile focused on the earliest phase—and found that the secret to snuffing out creativity among writers was to get them to think about getting rich and famous with their writing. "I found that writers wrote less creatively when you got them thinking about all of the external motivations for writing," Amabile explained in a dinner conversation we had while I was writing this book. "They did not do best when creating for others. They did best when creating for themselves." Creating what matters, what will endure when we are gone, starts with listening to our own experiences—and from these, creating things that over time we learn to whittle down to express elements of this experience that not only move us but eventually move many.

I had yet to learn this the year I lost my job.

Soon after my California fiasco, MIT professor Robert Langer invited me to join his group. Langer was a world leader in the field of drug delivery systems. I knew little of his field of research, and he knew little of mine. There didn't seem to be much we could do together, unless, of course, we did something completely new.

At the time, I had never thought about pioneering unexplored paths with a team that didn't know where it was headed, with such diverse expertise that the only thing we really shared was our innocence. I couldn't imagine how we could work together. In my first presentation to the Langer lab, I shared a hundred equations. Nobody understood. How else was I to express my ideas? When it came time for questions, nobody raised a hand. The blank looks seemed to question why I was here. I wondered it too. After it was over, I shared my concerns with Langer. "It doesn't matter what anyone else thinks, David," he said, laughing. "All that matters is what you *do*."

This was news to me. With Brenner, aesthetics had been about "form," meaning the shape and overall look of the things we created (formulas mostly). With Langer, it became about function, or what *came* out of those things we made.

He handed me three review articles he'd found that described the race to develop a way to inhale insulin for the treatment of diabetes. "Read up on it," he said. "Tell me if you see a better way."

Proteins—began one of the articles, starting with the basics—are made in the cells of our bodies. They diffuse out of these cells and go to other cells, tissues, and organs, where they react, produce changes, and get broken down. In other words, they are the natural therapeutics of our bodies. Having figured out how to make them in a lab in the 1970s, scientists had assumed the future of drugs would be protein based. But proteins, they'd discovered, were too big to effectively enter into your blood when you swallowed them. You needed hypodermic needles. There followed a race to deliver proteins into the human body without an injection. By the 90s, the scientific consensus was that inhalation beat all the other routes into the body that didn't break the skin. Once inside the lungs, a drug found a short, direct path into the bloodstream. When we breathed, our lungs pulled air into millions of little airways. Basic geometry showed that if you opened up all those airways and laid them out on the ground, the surface area of lung tissue would equal the size of a single's tennis court! So if you could find an efficient way to get proteins into the lungs, they would likely enter the bloodstream, and you wouldn't need injections at all. The prize application for all this was human insulin for treatment of diabetes. Insulin was small enough to cross the blood barrier with particular efficiency. If we could find a way to get insulin into the lungs with high probability, simplicity, and low cost, we might revolutionize diabetic health care. Two important efforts had been made by that time to put insulin in the air so that diabetics could breathe it into their bodies. Both worked, but they involved complex air-gun-like machines, and some wondered if they would ever make it through clinical development. Even if they did, it was unclear how many diabetics would actually use them.

The challenge intrigued me. I had little understanding of the problem and needed help. Langer had invited me to ignore others and follow my passion. I'd done that, and here I was with my mathematical expertise, at a frontier that excited me, only I didn't have all the skills and knowledge I would need to even have the chance to make a discovery. With Brenner, I had pioneered a less wild frontier. It had been one that I'd trained for, and my mentor had lived at this frontier for decades. The instincts he'd taught me, to pay attention to only what I knew or could discover from knowledge I had, would now be a disaster. I needed to open up and listen to others who did not always understand the value of this knowledge. I needed to care about what they cared about, read their literature, and change how I expressed my ideas, using language we could all understand.

In an odd gesture, Langer gave me literature to review that he'd never personally explored. Brenner had given me his own articles and asked me to advance an applied math field he had already pioneered. That was the traditional way—get to the frontier with the guidance of a mentor and then push it out a little further. Langer asked me not only to ignore his expertise, but to in fact ignore my own. "Go discover," seemed to be his motto. "Figure out something I don't know."

Before long, we discovered how to make an insulin particle you could inhale easily and inexpensively. It traveled efficiently into the lungs because it had the shape of a Wiffle ball. Mathematically, at least, it seemed you could get it into the air so easily you wouldn't need a complicated device, even a simple tube would suffice. It took two years to show it worked. Then in 1997 the results appeared in the journal *Science*. I expected this to dramatically change my life, but aside from a few days of euphoria, the article only raised more questions. Would my porous particle idea change anything? I started to ask around about ways to produce my particles less expensively and with other collaborators found simple ways to test whether they could actually deliver drugs to people as I'd proposed they would. This eventually led to a company that Bob Langer and I, with an investor named Terry McGuire, started in 1998 and sold in 1999 for over $100 million.

In retrospect, all this happened as a whirlwind. From an applied mathematician, who largely lived and worked (and frequently slept) in an office, to an entrepreneur, who haggled with lawyers and bankers as we worked out the sale of a company—I had embarked on a whole new life. I loved the newness, the stakes, and the opportunity of this off-campus life. I just couldn't figure out if this life actually belonged to *me*, in the sense of me-the-dreamer.

I bought an apartment in Paris with my French wife, Aurélie. We spent our summers there for a couple years while I continued to lead my company, now a subsidiary of Alkermes. This proved challenging. I couldn't turn my back on the company I had just sold, not even for a couple months in the summer.

We had developed the inhaled insulin product and started to imagine a new dopamine product for Parkinson's disease. Eli Lilly—a major US pharmaceutical company—had partnered with us on inhaled insulin, while inhaled dopamine, the product that would succeed in the marketplace fifteen years later, remained stranded. Lilly dominated the American insulin market with its injectable product. Relative to injections, the inhaled form demonstratively improved the lives of diabetics (to the degree that patients often resisted giving back our inhalers at the end of the clinical trials). For some patients, it was a life changer. But while our product succeeded in clinical trials, Eli Lilly eventually shelved it. This failure had little to do with the science—in fact, competitive inhaled insulins (the first-mover product

Exubera and later on Alfrezza) that made it to market demonstrated a problem we hadn't imagined. Medical doctors and insurers worried that the standard for insulin therapy needed to be set higher than the traditional FDA bar. By then one of every eight health care dollars went toward diabetes treatment. With prevalence of diabetes skyrocketing, doctors feared introducing an entirely new therapy. Insulin therapy was, after all, lifelong, and they had well over fifty years of experience with injected insulin: they knew it was safe. Replacing injected insulin with inhaled insulin, for which we might have a handful of years of experience, seemed to them a massive shift fraught with long-term worry. Beyond this, pharmaceutical companies, including Eli Lilly, had invested billions of dollars into insulin injections. Bottom line: there was little financial incentive, but significant medical risk, to replace needles with inhalers in the treatment of diabetics. So we had created a product that was going to eliminate a market that Eli Lilly already dominated to replace it with another product whose size and success was unproven. The risk of inhaled insulin beat out the reward it brought in improved patient lives.

Creating to matter was obviously more than a question of getting the science right.

In 2001, a chance came up to join the faculty of Engineering and Applied Sciences at Harvard University. The opportunity intrigued me. I'd left the university thinking I understood what mattered: newly discovered truths you could reproduce. You had to advance a good argument for why benefits followed from your discovery, which I'd done by writing textbooks. Off campus you actually had to realize the benefits. I'd discovered that important learning happened when you tried to bring benefit to people, and it involved neither completely deductive nor inductive creative processes but rather something in between. I wanted to explore where this learning belonged, if anywhere, on a university campus.

Once at Harvard I began to talk to faculty in architecture, music, and medicine whose work merged artistic and scientific processes. My conversations with colleagues expanded to include others outside the university, in museums, hospitals, performance halls. We differed in our backgrounds and passions, and yet we shared a common experience. Our institutional ties were tenuous and ambivalent, and we each liked to cross conceptual barriers in the pursuit of ideas of personal meaning (I had long written fiction as a personal vocation, mentored for a while by the MIT Writing Department and particularly by the novelist Anita Desai). Those who succeeded in creating things of durable value seemed to develop their creative aesthetic process in a public dialogue. I started thinking about how to bring creative work into a *culture lab*, where you listened and learned without millions of dollars and regulatory barriers separating you from the people you created for. What would it mean for me to open my lab to the public? Did I need sinks and centrifuges? Less turned out to be more. The lab needed to touch people, to give people fresh new experiences by way of the things we made.

In 2005 I moved my family to Paris, and in 2007 I opened up Le Laboratoire—an exhibition space built on one of the oldest corners of Paris, under a courtyard with massive stone pillars.

Le Laboratoire brought my work together with that of artists, designers, chefs, and perfumers in an experimental public conversation. Ideas emerged related to how we eat, how we deliver health care, and how we communicate in a world without sustainability. In 2014 I moved my culture lab to Cambridge, Massachusetts, added a restaurant, and opened it all in the middle of a neighborhood with the world's greatest density of technology companies.

Later we'll look at what came of (and out of) Le Laboratoire. We will also see how Le Lab belongs to the Grassroots Creator Movement that grew around the world over this same time frame and what it means to the way each of us learns, works, and lives—giving us the chance to contribute to a hopeful future.

But first let's look more closely at this idea of the culture lab.

Figure 1. A first artwork of Le Laboratoire, *Singing Cloud* (2009) by the Indian artist Shilpa Gupta, is now in the permanent collection of the Louisiana Museum in Denmark.

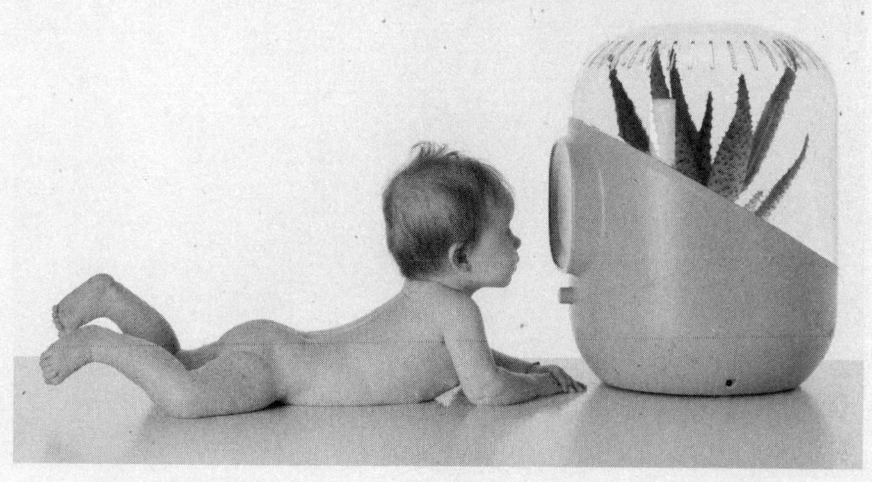

Figure 2. The first commercial design of Le Laboratoire, *Andrea* was created by the French designer Mathieu Lehanneur in collaboration with the author and is now in the permanent collection of MoMA in New York City and the Musée des Arts Décoratifs in Paris.

Figure 3. *The Refusal of Time*, created by William Kentridge in collaboration with the historian of science Peter Galison and first exhibited at Le Laboratoire in 2011, became a video installation, and eventually led to the critically acclaimed theatrical companion *Refuse the Hour.*

Figure 4. Café ArtScience in Cambridge, Massachusetts, conceived by the author with interior design by Mathieu Lehanneur, and opened in November 2014.

Aesthetic Dimensions and the Culture Lab

Culture labs, like Le Lab, are settings that support the experimental creative process from idea conception to realization. They can take all sorts of forms, from experimental restaurants, websites, startup companies, and independent theater companies to science, engineering, and design labs. Culture labs promote mindful engagement and public learning. Unlike large organizations—with leadership more adapted to adhering to rules than to ambiguity, and where failure is not an option—culture labs remove obstacles to change. They tend to be places of hopefulness, for the dreamers who work inside them and for the public that participates in their dreams.

Culture labs have just a few defining rules. There is a team of two or more people who develop a creative idea and exhibit it to the public at multiple points on the path to realization. There is a leader—the writer, the inventor, the chef—and maybe more than one (as in the Beatles or the Apollo 11 team). The structure of the team can change as the idea changes. Activators ranging from investors to early audiences, technicians, assistants, and others, including artists, designers, and engineers, guide the team's ability to pitch what they have learned and learn from what they hear. Pitches, which express the first impression of a creation, are not the formatted presentations that might be made in the invitation to invest in a new commercial startup company. They express a creative idea that has never been imagined before, whose utility is impossible to know. A perfect culture lab pitch is literally a work of art.

There are two kinds of culture labs. Transient labs, most familiar to us, pop up for a short time—a weekend, a month, a semester—and frame the creative process, often with little or no aspiration to last. Aspirational labs, like those I describe in the next chapters, frame the dreams of artists, scientists, chefs, and other creators over long periods of time. They require support and guidance and are critical to creating the future. Transient or aspirational, culture labs may be public learning institutions, like Le Lab, or they may be private commercial operations, like a restaurant.

In culture labs, we create new things with and for others while bearing the real risk that these new things will fail to please. Why do we take this risk? The answer goes beyond our personal motivations to create a song, a gadget, a home, any of which may please us and might even earn us praise or money. Creation stirs within us the profound biological sense that we are not alone, that we are wired for relationships, and that we rely on each other for our mutual well-being.

Contemporary research in neuroscience points to particular emotive and cognitive states that favor creating for collective benefit and for personal wellness gained through the process. Fiercely self-centered, our creative brains turn out also to be social.

If we laid out the human nervous system on a table, we would notice two very different-looking structures. One is the lettuce-like structure we identify as the brain, with its undulating cortex, the gray matter of our brains, the central nervous system. The other is a network of fibers, the peripheral nervous system, the family of nerves that stretch out through our bodies and gather signals, from within and without, and hustle them back to the brain. The primary sensations of sight, hearing, smell, touch, and taste transmit via our nerves from the world around us to our brains. Other signals, like hunger, pain, or anxiety, originate from inside our bodies. Our brains process these signals into actions and feelings, promoting an impression of where we are, how we are, and eventually who we are.

Particular emotive and cognitive neurological states help us survive in pioneering circumstances. These "aesthetic dimensions" of the creative mind help us engage with our environment and bring benefit to ourselves and to those around us.

In the stories of artists and scientists that follow, I highlight seven such aesthetic dimensions. Here is a first look:

Passion. Marie Curie did her doctoral research in Paris in the late 1890s on the seemingly magical properties of uranium. Others had shown by then that uranium generated x-rays all by itself but the reason remained a mystery. The uranium phenomenon implied danger. Curie was passionately curious to figure out where the x-rays came from and began to play with the material, manipulating it and assessing it as if the long hours she spent with uranium would not shorten her life. Within a decade, she published her discovery of radioactivity and won two Nobel Prizes.

Creating things that matter starts with a passionate curiosity to discover. This passion rarely lets up over the long course of pioneering exploration and leads to a created thing that brings fresh experience to many others.

Empathy. We'd best not climb a peak or dive into the ocean without someone at our side. The probability of our survival goes up when someone else is thinking about our interests. Survival experience builds a reliance on another that gives a pioneer confidence that in the most difficult and unexpected of circumstances, she is not alone. She ventures further than she might otherwise have, as Marie Curie did in collaborating with her husband, Pierre, and winning with him her first Nobel Prize; or John Lennon, in creating an iconic band with a group of friends; or John Glenn, in pioneering lunar exploration with the Apollo 11 team.

By teaming up at risky frontiers in the empathetic way of successful pioneers, discovery of what passionately interests us turns into the discovery of what matters to others.

Intuition. Our brains accumulate over time deep experiential knowledge that creators know to use by placing themselves in circumstances that are cognitively new while emotively familiar. In the absence of specialized knowledge, precious unconsciously stored information comes to them, which they intuitively act on, and with great surety, as when Mark Zuckerberg (long before he became CEO of one of the most powerful companies on earth) created Facebook as a Harvard sophomore in 2003.

A year before he had the idea of Facebook, Zuckerberg caused a scandal by putting photos of Harvard students onto a website that compared faces and asked students to vote on the more beautiful one. The site, called Facemash, was briefly popular, but it irked many on campus, and Zuckerberg had to take it down. Shaped by the Facemash experience and other social network experiences going on around campus at the time, Zuckerberg intuitively and aggressively pulled his new project together from a nascent dating site on campus. A year later, Facebook activity exploded.

Innocence. We tend to be most in tune with our intuition and notice things around us more when we are innocent to our circumstances than when we are deeply knowledgeable of them. Innocence propels learning, and passionate, empathetic creators use it to learn fast, pioneering out of their innocence and benefiting by direct personal experience that is now grounded in the way things actually are.

John Cage was thirty-nine years old when he stepped into an anechoic chamber at Harvard University. The brilliant composer of silence discovered the "sound of silence" in a sound-absorbing (anechoic) chamber. "I heard two sounds," he wrote in his work *Indeterminacy*, "one high and one low. Afterward I asked an engineer in charge why, if the room was completely silent, I had heard two sounds. He said, describe them. I did. He said, the high one is your central nervous system in operation. The low one is your blood circulation." Neither explanation was true—the sounds he heard definitely came from the functions of his body, while tracing those sounds to the functions themselves was a leap of imagination. Cage had no training in physiology. Nor did he need it. His innocence promoted an understanding that went beyond physiology and foreshadowed one of the greatest musical creative careers of the twentieth century.

Humility. Creators of very new things know they are often wrong. They learn more through making hypotheses that prove incorrect than by their successes, which at best confirm what they already know. Although they may be some of the most successful people, lifelong creators possess a humility that associates with their willingness to be passionate, empathetic, intuitive, and innocent long after many of us "grow up."

Steve Jobs, Bill Gates and Paul Allen, and Michael Dell created a few of the most dominant companies of their times by frequently learning from what did *not* interest people—in product failures large and small—how to make new technology that *did*. Jobs failed to market his NeXT computer and learned from it in reviving the Macintosh; Bill Gates and Paul Allen failed with their first startup, Traf-o-Data, and learned from it in their creating of Microsoft; and Michael Dell, who had failed to grow his company into the notebook era, changed his management in 1993 and built one of the most successful computer companies of the 1990s.

Aesthetic intelligence. Creators who express new ideas do so with a deep understanding of the power of form, whether in words, equations, or musical notes. They express their ideas in forms that attract the interest of those who are in a sense first adopters and will enter into a conversation about aesthetic value. Even more important than the brilliant function of new ideas is the form that makes them beautiful to experience.

By the time Albert Einstein published his first article on the theory of relativity in 1907, with little public fanfare, Pablo Picasso had created a public sensation with his *Les Demoiselles d'Avignon*, painted two years before. *Demoiselles* portrayed the modern sense of the mutability of space and time, which Einstein could feel in Vienna as Picasso did in Paris, two twenty-somethings living in a Europe whose space and time had been transformed by train travel.

Obsession. Doing anything new that impacts how others live is to meet resistance from those who, for whatever reason, fear change and those who might lose opportunity as a consequence of the arrival of the new creation. Pioneers who bring very new creations to life tend to so deeply associate their own lives with the existence of what they have made that they commit to the survival of their creation as if it were their own survival.

Einstein's theory of relativity, which postulated the incompleteness of Newtonian physics and a vision of physical reality opposed to our intuitive sense of the world, met strong resistance. For scientific, theological, and philosophical reasons, people disputed Einstein's new view of the universe. Throughout his life, Einstein deepened his understanding of the theory of relativity, ever resisted by a dwindling core of naysayers, while increasingly supported by scientists, who gradually confirmed its basic predictions.

These same aesthetic dimensions propel leading creators today and are as fundamental to the Grassroots Creator Movement as they are to the future of learning. In the next chapters, I tell the stories of nine amazing creators working today in the arts and sciences to describe how we too can find personal fulfillment in the creation of things that last, through the nourishing of these emotive and cognitive states inside our own culture labs.

THE CREATOR'S CYCLE

CHAPTER 3

Ideation: What If I Don't Even Have an Idea?

A few years ago, I stood in front of forty or fifty students at the start of the spring semester at Harvard University, fielding questions. I had just shared the basic content of my spring course, How to Create Things & Have Them Matter, and explained how we would be approaching the coursework over the semester. In our next class, I planned to share four dreams. My dream hypotheses would be vague, even outlandish. "Might we breathe food?" appeared on my slide. "People have for a very long time used forks and knives, or chopsticks, and for an even longer time, fingers. What if we could simply breathe food into our mouths? What would this mean to human health and wellness?" Ideas like this one were designed to make my students dream without assigning a problem to solve. I explained that in our next class they would gather around a chalkboard and list their names under two of the dreams that most interested them. We would then form groups of four to six students per idea, and if one idea proved unpopular, we would brainstorm a way to turn a more popular idea into two ideas.

The semester had three phases. In the first phase, they were to work as a team to create their *own* idea that would address a major need or opportunity. My ideas were like signposts that pointed students to a frontier. In the second phase of the class, each group would express their idea, explain how others have tried to meet this need or opportunity, and imagine a first simple experiment, which they might test during the summer. In the last phase of the semester, they would develop their pitches, speak to experts, and finally deliver the pitch in a public forum with a final report they would write as a team. Half their grade related to the quality of the final pitch and report, and half related to participation during the semester. Thanks to financial support from the Wyss Institute of Biologically Inspired Engineering, to which I belonged, they could win grants to develop their ideas over the summer if they felt particularly passionate.

I felt I had been clear, and the questions I'd fielded so far confirmed the feeling. The students were in the process of putting together their spring semesters, and this was "shopping" day. They might take the class or not.

A young woman seated in the first row raised her hand. Her face seemed strained.

"What if I don't *have* an idea?" she asked.

Her question at first seemed funny to me. How could she not have an idea? I gave it a second thought. Yes, it would be a disaster to take a course at Harvard University with no chance of doing well because she simply couldn't come up with an idea. It was a little troubling I'd never considered it. Why hadn't I? I'd lived more or less in the hypothetical way proposed by the course I was now teaching. I couldn't recall ever thinking that I just wouldn't be able to come up with an idea. After a few seconds of silence, I laughed. Obviously, she had a much higher standard for her ideas than I had for mine, I said. My students now laughed with me. But I was serious. She was a Harvard student, and she couldn't imagine an idea that was less than brilliant. What other kind of idea would she have?

"I maybe have a hundred ideas every day," I said. "Most are mediocre, some are just bad, a few are perhaps good. I don't spend much time trying to figure out which is which. Actually I'm not sure I even could. So I try to hang around people who care about ideas and who feel free enough to share them unquestioningly. Over time, we play a kind of idea ping-pong. This class is about playing idea ball. It's a fun game. And if you win the game—which, by the way, nobody needs to lose—you have the chance to create your own futures."

Coming up with an idea that matters to anyone else can seem like a daunting challenge. How can we be sure that an idea is worthwhile?

Having a creative idea and working to realize it is about starting and carrying on a passionate conversation that kicks off with curiosity and accelerates with a team bound together by empathy.

How do we cultivate passionate curiosity and empathy? And how can we learn to use these qualities as effective creators? Some examples of those who are famously creating in this third way can help us begin to understand how this might work in our own lives. These creators are not driven principally by a thirst for financial gain or necessarily a will to achieve cultural impact, but by a desire to live the aesthetic life of the pioneer. They work inside culture labs that shape the creative process and develop a creative dialogue with the public leading to created things that matter.

Reimagining How We Eat—Aesthetic Passion in Action

Ferran Adrià is the world's most celebrated living chef. At El Bulli, a tiny restaurant in the foothills of Roses on Spain's Costa Brava, Adrià snapped the four-century French dominance of exquisite high, or haute, cuisine. By a meticulous personal interpretation of the food culture of his native Catalonia, Adrià attracted popular attention as no chef ever had and ignited the local food movement that is at the center of the public conversation around the future of food.

Before Adrià, haute cuisine meant the French approach to food. In bourgeois France, one did not actually speak of eating. One spoke only of lunch and dinner, as if the experience of the French table was one thing, and what people did with food elsewhere was something else. French sophistication set the high bar of what eating could be, if one worked hard at it.

The French introduced haute cuisine in the seventeenth century. The royal chefs La Varenne and Massialot, creators of staples like marinades and ragouts (stews), published the first cookbooks. In the late eighteenth and early nineteenth century, Marie-Antoine Carême, the famed chef of the French diplomat Charles Talleyrand-Périgord, Czar Alexander I, and the banking scion James Rothschild, brought haute cuisine post-Revolution to bourgeois tables, creating fabulous sauces (Carême had around a hundred in his kitchen) such as velouté and béchamel. These entered the haute cuisine treasure chest and remain there today. French cuisine continued to pile up treasures, with a string of famous chefs, like George Auguste Escoffier, who opened the Ritz and Carlton hotels (the two later merging to form the famous Ritz-Carlton hotel chain) and invented new recipes like Peach Melba (created for the Australian singer Nellie Melba in 1893). Charles Ranhofer brought the French tradition to America as the chef for thirty-four years of the New York restaurant Delmonico's, where he created Lobster Newburg and many other dishes. After Ranhofer came America's great chef Julia Child, who wrote *Mastering the Art of French Cuisine* and became the first massively popular television chef.

Eventually, it became hard to imagine, at least in the West, anything appearing on a proper table that could compete with the refinement and sophistication that four centuries of cooking gave to French cuisine. Not many even tried.

Then came Ferran Adrià who, over a twenty-five year span from the mid-1980s, changed the trajectory of haute cuisine, making it accessible to anyone—from a remote coast in Spain. Before Adrià, brilliant artistic minds turned to careers in design, advertising, and the arts. After Adrià, many wanted to become chefs. Creative cuisine started to show up in reality television, popular movies, and competitions. Eating well became a creative engagement and as it did, the agenda of eating moved in the direction of sustainability, biodiversity, and the future of food.

Interestingly, Ferran Adrià—a small, compact, powerful-eyed Catalan, the most intensely curious and passionate person you will ever meet—did not start out passionate about food. "I grew up in L'Hospitalet de Llobregat, a poor village outside of Barcelona," he explained to me over dinner one night in the winter of 2016. He had invited me to his brother Albert's Mexican restaurant, Hoja Santa, near Barcelona's Plaça d'Espanya. We spoke French. Adrià understands English but refuses to learn to speak it. All his public speeches, and of course his work, are conducted in Spanish or Catalan, while his facility in French seems to reflect the pilgrimage it took to bring the world's culinary zeitgeist across the border to Spain. "My mother was not a good cook. She made food like any other mother in our village did: without thinking too carefully about it." A server arrived and placed a small plate in front of Adrià. It held corn bread shaped in the delicate form of a small cornstalk. "Food was who we were, the heart of our culture." The great chef placed salsa on the back of his hand and quickly licked it off. "We didn't see it because it was all we knew, and because we didn't see it, nobody I knew made much effort to create it with care."

Adrià played soccer, eventually tired of school, and found a job through a friend of his father (a plasterer) working in a local beachside restaurant. With money in his pocket, Adrià dropped out of school and, for the first time in his life, started to learn his way around a kitchen. A year later, when he moved to Cartagena for his compulsory military service, Adrià began to cook for a Spanish military officer instead of participating in standard military drills and service. There he made a friend, Fermí Puig, who was also doing his military service by cooking meals for the Spanish officer. The next summer, Puig, today a celebrated Barcelona chef, convinced Adrià to do a summer internship at El Bulli, at the time a quaint restaurant on the Costa Brava set up years before by a German couple named Schilling.

Built on the edge of a bay in Cala Montjoi, a scenic bike ride from Roses, El Bulli catered to summer vacationers, who would sometimes drive their boats up to the coastline, drop anchor, and raft in for dinner. In the 1970s, the restaurant had adopted a French model and earned its first Michelin star. A second Michelin star came with French chef Jean-Paul Vinay, who moved from France to Cala Montjoi a few years before Adrià arrived. Adrià interned with Vinay, sharpened his French, and after completing his military service came back to the restaurant full time. By then Vinay had approached the Schillings looking to buy them out. The Schillings refused, and Vinay abruptly decided to move on.

Adrià quietly took over the kitchen.

Worried about the future of their restaurant and recognizing the primacy of the French touch in haute cuisine, the Schillings sent Adrià—chef de cuisine at the age of twenty-five, basially innocent of French culinary culture—north of the border during the winter season when the restaurant was closed. In 1987 they sent Adrià to Nice to attend a demonstration by Chef Jacques Maximin at Le Chantecler in the Hotel Negresco. Maximin had by then the distinction of being the first chef to receive two Michelin stars inside a hotel palace. Rumors were that he might have earned three had he not been responsible for room service. The superstar chef completely changed his menu from one year to the next. Adrià had never seen anything like it. Every recipe on the Chantecler menu belonged to Maximin. He created things. But did he really? Adrià wasn't sure. Maximin's food stood at the center of French tradition, with its ragouts and marinades, its parsley sauces and truffle mousses.

"What does it mean to create?" someone suddenly asked.

"To create is to not copy," replied Maximin.

Copying was in fact all that Adrià did at El Bulli. He had come to Nice to learn more recipes to copy. Copying masters of French cuisine wasn't easy for Adrià, who had come from a suburb of Barcelona and a family in which cooking had essentially been routine. In this light, Maximin's comment was a liberation for Adrià. It was the easiest thing in the world for Adrià to be original! He needed only to draw inspiration from where he came.

Becoming a great chef like Maximin apparently meant discovering himself—and expressing his Catalan experience in food that others, who did not share his background and experiences, would find beautiful.

Adrià came back from Nice and immediately tossed

out his nouvelle cuisine handbooks. He began to reimagine classic Catalan sauces, shaping his vision with ingredients that had never been tried on haute cuisine menus before, like crayfish roe and sardines. He naturally adopted the vernacular of the fanciful and subversive tradition of the brilliant Catalan masters, like the otherworldly architect Antoni Gaudí or the surrealist artists Joan Miró and Salvador Dalí. As he advanced, his kitchen grew with an enthusiastic Catalan family. By the late 1980s, it included his brother, Albert, who had quit school at the age of fifteen to come work in the kitchen of his older brother—soon to become one of the top chefs de patisserie on the planet. José Andrés joined his kitchen as well. Andrés was also a teenager when he began and later became a pioneer of small plate cuisine in the United States.

The El Bulli family worked hard, while its members also partied hard in Barcelona discos ("Nobody at El Bulli expected me to last that long as chef de cuisine," Adrià says with a laugh). Over time the Adrià team intuitively pioneered what later became known as modernist cuisine.

As the 80s slipped into the 90s, the new chef at El Bulli started to catch attention. In 1990, the restaurant won back the second Michelin star it had lost with the departure of Vinay, Adrià bought out the Schillings, and the chef started to lift his eyes to his future horizon. He spent a winter in the Barcelona workshop of his sculptor friend Xavier Medina-Campeny, cooking lunch and dinner and absorbing the spirit of an artist's studio. The next year, he renovated the kitchen at El Bulli. By 1993, his tapas were a local sensation. His menu that year included black truffle water ice with celery jelly, mozzarella with bacon, dried apricots with mustard and caramel vinaigrette, and sardines with cauliflower couscous and ragout of cauliflower pulp and pig's ear. Young Catalan men had by now begun the habit of bringing their girlfriends from Barcelona, traveling up the winding mountain road to the little restaurant by the sea and watching their girlfriends weep with joy at cuisine more delicious than they'd ever imagined.

Science was entering the haute cuisine kitchen in those years. In the late 1980s, Oxford University physicist Nicholas Kurti and French physical chemist Hervé This had started holding an annual "molecular and physical cuisine" workshop in Erice, Italy, where they attracted renowned chefs like Heston Blumenthal and Pierre Gagnaire. Adrià points out that he never participated in the conference and had no scientist working at his side until the mid-2000s. Some of his inventions of the 1990s, common to the repertoire of *molecular cuisine*, came about independently through intuition and trial and error.

For instance, French mousses had a long tradition in French haute cuisine. Traditionally, you made them by beating liquid preparations with a fork. Depending on the nature of the instrument you used to beat air into the liquid, and your ability to beat hard and fast, the mousse could have a range of airiness. In the early 1990s, Adrià dreamed of an extremely airy mousse. During his winter in the workshop of his sculptor friend Medina-Campeny in 1992, he came upon an oxygen tank. It occurred to him that he might simply point the oxygen into a tomato. So he connected the free end of a tube attached to the oxygen tank and opened the nozzle. A few seconds later, bits of tomato burst into the air. He repeated the experiment but never managed to keep the tomato from shattering. The next year, a friend, the pastry chef Antoni Escribà, showed him a carbon dioxide siphon from Switzerland, which got Adrià thinking. Soon afterward he discovered a nitrous oxide siphon at a nearby restaurant called Bel-Air. With his brother, Albert, Adrià borrowed the siphon and created his first food foam, or *espuma*. Gelatin seemed to be the key to making the foam retain its structure even with all the air inside. That season he introduced his first *espuma* onto the menu of El Bulli, an extraordinary foam of white bean on top of black sea urchin. Airy mousses became a signature of Adria's cuisine.

Adrià and his team built the El Bulli legend across the 1990s as the Internet came to life. He seemed to bring new ideas into his restaurant as quickly as the Internet brought new ideas into every home.

Adrià had an idea. He experimented. He served. This pattern also reflects how the human brain processes sensory data and responds with creative action.

The brain's motivation to create stems from a sophisticated reward system that involves the release of neurotransmitters such as dopamine and somatostatin, the gathering of sensory information from inside and outside the body, and the activation of neural structures in the basal brain (specialized for muscle function, balance, and autonomic behavior such as breathing), the limbic system (the middle part of the brain, specialized for memory and emotions), and the neocortex (the primary mass of the human brain, specialized for higher cognitive functions). Scientists divide this cognitive-emotive reward system into three basic functions—wanting ("incentive salience" or motivation), associative learning (positive reinforcement), and liking (including joy and ecstasy). These three reward-system functions aid innovators in the creative process from conception to realization and promote passionate curiosity.

We want or desire certain things, like shoes, because they give us pleasure as a direct benefit (they feel good on our feet, we avoid foot injuries). We want other things, like money, because they deliver pleasure indirectly by representing value and future benefit, giving us the ability to acquire things, like new shoes. Our brains learn about indirect benefit through reinforcement learning, which can take what scientists call a "model-free" form and a "model-based" form (more on this later), each involving different but overlapping circuits. A student may initially want to read *War and Peace* to get a good grade in her Russian literature course, largely because her brain possesses an abstract "model" that getting a good grade will reap future benefits. Meanwhile, the process of reading Tolstoy, brushing up on French and Russian history, and eventually becoming transported to a nineteenth-century world of emancipation may prove so immediately pleasurable that she develops a lifelong love of reading world literature. She now *likes* to read.

Wanting, reinforcement learning, and liking derive from signals that pass through our nervous system, leading to reflections, decisions, and actions. Such signals and processes overlay and participate in the complex cognitive and emotive functions of the brain during the entire arc of the creative process. They motivate us to dream up, experimentally develop, and express new ideas—or what I call the Creator's Cycle: ideation, experimentation, and exhibition.

Through the repetitive experience of the Creator's Cycle, we maintain and sharpen the cognitive and emotive dimensions of the aesthetic creator life, starting with passionate curiosity.

Adrià's passion grew in the El Bulli kitchen. Step one, *he ideated or dreamed*, throwing away his books on nouvelle cuisine and returning to the food of his native Catalonia. Step two, *he experimented with others to realize his dream*, surrounding himself with a Catalan team, working relentlessly in his studio and with artist friends like Medina-Campeny. Step three, *he expressed his dream to others* in every plate he served, never repeating himself, as he was a creator and, according to Jacques Maximin, to create meant "to not copy."

In August 2003, a photograph of Adrià appeared on the cover of the *New York Times Magazine* with the title "The Nueva Nouvelle Cuisine: How Spain Became the New France." In the cover article, Arthur Lubow hailed Adrià's restaurant as "a gastronome's once-before-you-die Mecca." He had visited El Bulli during the restaurant's twentieth-anniversary season, which Adrià celebrated by serving some of his greatest hits. Lubow wrote, "The menu included 30 tapas-size dishes, each identified by the date of its introduction. . . . Welcoming cocktails of a frozen whisky sour and a foam mojito were accompanied by popcorn that had been powdered and reconstituted as kernels and a tempura of rose petals. The Catalan mainstay of pa amb tomàquet, which is grilled bread rubbed with tomato and drizzled with olive oil, was deconstructed into a white sorbet made from skinned tomatoes and topped with a dry cracker that was filled with olive oil. A chicken croquette contained liquid consommé. A 'Kellogg's paella' consisted of puffed Rice Krispies, to which the waiter added an intense seafood reduction; on the side were a small, flash-fried shrimp, a piece of shrimp sashimi and an ampule containing a thick brown extract of shrimp heads that you were instructed to squeeze into your mouth."

Before the *New York Times Magazine* article, Adrià's restaurant had sometimes remained empty through an evening. After the article came out, and until he closed his restaurant, Adrià would receive each year over a million requests for a reservation at El Bulli. His restaurant became to haute cuisine what the Internet had become to media—a disruption to a long-standing and successful way of getting to people things they wanted to consume. El Bulli was chosen by *Restaurant Magazine* in 2002 as the best on the planet. Eventually, it would win the award a record four times in a row before closing down in 2011.

After he became famous, Adrià might have taken fewer risks. Fame can distract innovators from the intrinsic personal rewards of the Creator's Cycle. They turn conservative as a consequence, anxious to avoid failure. Adrià loved what he did too much and continued on as he had since he first arrived at El Bulli. His relentless commitment to never repeating himself made El Bulli unique in history, but it also exhausted him as public pressure grew to deliver perfection to every customer.

Adrià's culture lab had turned into a place where, improbably, the three ways of creating mixed together. Not only did it encourage the third way of dreamy creation, it produced a cultural product every day (the "second way") and was at the same time a profitable commercial company (the "first way"). Lifelong aesthetic creators tend to develop and hone their aesthetic process in culture labs that are mostly dedicated to the third way of creating, protected from the challenges of production. They figure out how to hand a manuscript off to a publisher or spin a company out of a lab and go back to the writing and inventing that fires their passionate curiosity.

Adrià had been alone at his frontier and without competition. After the *New York Times Magazine* article, his frontier became the hubbub of the haute cuisine marketplace. Competitors now existed all over the planet. Critics expected him to retain his number-one status year after year. To Adrià, his culture lab lost its frontier feel, and this made El Bulli unsustainable. After several years of unparalleled success, he closed it down.

Closing El Bulli did to haute cuisine what closing the most famous art school in history, Germany's Bauhaus, did to the study of art, design, and architecture. When the Bauhaus closed in 1933, its geniuses traveled abroad like seeds in the wind to Chicago (via Mies van der Rohe at the Illinois Institute of Technology and Lázsló Moholy-Nagy at the New Bauhaus, now IIT Institute of Design), to North Carolina (via Josef Albers at Black Mountain College), and to Boston (via Walter Gropius at the Harvard Graduate School of Design). After El Bulli closed in 2011, the top restaurant award passed on to chefs who had trained with or been deeply influenced by Adrià, including René Redzepi at Noma in Copenhagen, the Roca brothers in Barcelona, and Massimo Bottura at Osteria Francescana in Modena, Italy. Adrià's influence quickly spread across Europe and the Americas, highlighted and formalized by Nathan Myhrvold's exhaustive 2,438-page cookbook *Modernist Cuisine*, which appeared in 2011, the year of El Bulli's closure. The new rules of the futurist cuisine movement read as if Adrià had made them all on his own.

I first met Adrià in the fall of 2015 when he walked into my restaurant, Café ArtScience, at the end of an event organized by MIT. I was in the early days of creating my digital scent platform and had designed a digital scent menu for the dinner. Each course had a scent that the menu, displayed on a digital tablet, emitted from the edge of the protective tablet cover with a touch of your finger. Adrià played along with the experimentation. His nose practically touched every food form we served him—the fish, the potato, the wine. When the dinner was over, he called out the chef, Patrick Campbell, and patted his stomach with the affection of an old-world father congratulating a son. Then he gave the talk I would hear him deliver every time I visited him in Barcelona. "What is this?" He held up an empty wineglass, while business partner Lluís García translated Adrià's words into English. "A wineglass? Maybe it's a dome for aroma." He turned the glass upside down and made us understand he might fill it with smoke. "A carafe? A bowl? A vase?"

Adrià's embrace of the experimental, his invitation that we question everything, was so remarkable that Nobel Laureate Phillip Sharp, who was there with us at the dinner, described the evening in an email the following morning as "the most memorable dinner I have ever experienced."

A few months later, I traveled to Barcelona to visit El Bulli Lab, which Adrià created in the center of Barcelona after closing down his restaurant. Interestingly, El Bulli Lab has no obvious kitchen. Twenty or thirty people, mostly volunteers from around the world, sit at tables quietly working. Surrounding them are posters and digital screens, books of photographs, and culinary artifacts. They belong to Adrià's ambitious project of quantifying and cataloging his original view of food, which is being published as a series of books called *BulliPedia*. At one end of the large warehouse-like studio is an exhibition of his project with Dom Pérignon to help the classic champagne brand reimagine itself. At the opposite end is a large interactive digital screen that allows Adrià to demonstrate his plans for reopening El Bulli on its original site as a cross between museum, school, and restaurant. Adrià's ultimate culture lab will be called El Bulli 1846, which is the number of original dishes the great chef made over the lifetime of his restaurant.

I spent two days with the chef and returned the next year for another three. On my visits, I followed Adrià around wherever he went, studying him as he looked over the shoulder of team members at work on *Bulli-Pedia* or as he strolled through a wholesale food market in Barcelona discussing food classification systems with local buyers. His facial expressions can seem clownish. If a collaborator isn't paying attention, Adrià might throw a horrified glance of disapprobation. Then his look might become analytical as he inspects a purée with surgeon-like resolve. He never seems bored or confused or anxious in a nonproductive way. Even in moments of reflection, he will engage the attention of those around him, as the shape of his body—the tense way he leans forward over a counter or listens with his neck purposefully inclined—somehow warns that he is about to challenge assumptions, which he inevitably does.

Aesthetic passion is the enduring conviction that our curiosity matters. Believing we matter is essential to creating what will. Over time, the created thing can seem indissociable from the creator himself. Creator passion can also be blinding if its only goal is personal gain. When we wish only to win a game, an argument, or a war, we are sure of who we are and of what we need, so in our search for success we learn less, change less. Aesthetic passion is not like this. It is the natural inclination to grow, within the constraints of place and time, as if we were a plant stretching out new roots looking for nutrients, adapting to water in the soil, to the angle of the sun, and to the force of the wind.

Ferran Adrià creates this way.

Reimagining How We Get Our Meds—Aesthetic Empathy in Action

In search of discovery at any frontier, whether it be literal, as with space travel, or figurative, as with the intellectual pursuit of the future of medicine, pioneers tend to bond with those at their side. Empathy in the extreme conditions of frontier survival is an essential dimension of creating to matter, as illustrated by the life and career of MIT professor Robert Langer, one of the most prolific inventors and the most cited engineer in history.

Early in his career, Langer created a biodegradable particle, generally invisible to the naked eye, to release therapeutic substances into biological tissue evenly and gradually while it erodes. His particle has led to new cancer therapies that have saved lives and broadened access to health care at a time when this access is narrowing.

Langer began his cancer work right after getting his doctorate at MIT in chemical engineering. He had grown up in Albany, where his dad ran a liquor store. He loved to teach—so much that he created and taught a chemistry program for high school dropouts at the Group School in Cambridge while getting his PhD. "You see this light go on in their heads," Langer explained to me while describing those days. "They just learned something about the way the world works, and it gives them hope they didn't have." Teaching for Langer was even more about "the light" going on than the sharing of the basics of chemistry. This passion to light up the world became central to what later made Langer a creator of things that matter to us all.

He tried hard to land a teaching position after graduating from MIT, and when he couldn't, he found an unlikely mentor. Judah Folkman was a renowned Harvard Medical School professor. Tall, poised, and confident in the way of the remarkable surgeon he was, Folkman had attained the position of chief of surgery at the youngest age in the history of Boston Children's Hospital. As a medical doctor, he was unquestionably brilliant. But his colleagues sometimes ridiculed his overly intuitive approach to cancer research. He believed that tumors grew in the body by sending out molecules as "signals" to blood vessels that then grew in their direction and eventually kept them alive. He called the phenomenon angiogenesis. His fascinating hypothesis was that angiogenesis could be blocked with imagined molecules that would eventually be called statins—angiogenesis inhibitors. Statins were going to kill cancer. Folkman had demonstrated the phenomenon in animal experiments, but isolating the specific factors involved was a challenge, and the biochemistry of angiogenesis remained unclear.

Langer came into Folkman's lab as the only engineer. It was a funny pairing, the quantitative engineer with the intuitive surgeon. Going into the Folkman lab closed the door on the many industrial jobs Langer might have taken after graduating with his PhD from MIT. Langer's decision had very little precedent in academia. His advisors back at MIT actually called it suicide. He didn't know enough to say if they were right. He also didn't much care. Important unresolved problems seemed to be everywhere in the Folkman lab. And none was as big and ripe as the problem of angiogenesis. How to prove angiogenesis actually happened?

Langer turned to young medical doctors working in the lab, like the intern Henry Brem, with whom Langer would coauthor one of his first papers. (Later in their careers, they would create a new FDA-approved brain tumor therapy.) Brem was tall, affable, and compassionate, in all these ways like Folkman. The son of Auschwitz and Buchenwald survivors, Brem shared with Langer and Folkman a belief that one should try to make the world a better place.

The leap Langer took going from MIT to Harvard Medical School was gigantic. Up until then, success had been about learning. Once he'd settled in Folkman's lab, Langer saw how he could create things, not just with what he knew, but often with what he didn't know.

By his own admission, Langer had not been a particularly creative kid. He remembered doing the things a typical kid did growing up in Albany in the 1960s and working hard at his studies. He had always seen his postgraduate options as either teaching kids to help them "turn on the light," or applying the engineering he'd learned over the years. But having failed to land a teaching job, he'd discovered that his second choice of "becoming an engineer" held little interest for him. He'd actually received twenty offers from oil and chemical companies upon graduating from MIT and had refused them all. Folkman offered an exciting new option. Creating new cancer therapies was a matter of applying logic, but it was also about mind-sets that counted in the experience of an effective teacher, including intuition, imagination, ambiguity, and uncertainty.

As a scientist, Folkman was frequently criticized for his overly intuitive nonscientific or artistic way. Interestingly, Langer saw that it was difficult to discover anything without such an open-eyed perspective. Obviously, as a scientist you followed logic. But with the surprising conditions you encountered at frontiers, relying on logic alone could make you miss the unexpected.

Soon after arriving in the Folkman lab, Langer hazarded a wild hypothesis. The large molecules that Folkman believed inhibited angiogenesis could be stored in a polymer particle. Polymers are the large complex molecules that make up synthetic plastic and also much of the natural tissue and biochemical matter of our bodies. Natural polymers like proteins diffuse through and get stored in the tissues of our bodies, where they do useful things like aid in wound repair and help tissues grow. Langer imagined creating his own man-made polymer particles to do the same thing. He would place them near the tumor and over time, the particles would release the large statin to kill the cancer without a doctor having to inject it all the time, eliminating hospital visits and giving those without access to visits a chance to live. Researchers like Folkman knew the idea Langer had in mind worked for small molecules. But the complexity and fragility of large statin molecules made everyone believe it would never work for Folkman's dream cancer therapy. The large molecule would lose its all-important shape or would eventually break down chemically and lose its efficacy. Langer, of course, had no proof that he was right. He couldn't deduce it logically, as indeed there were plenty of ways for the large molecules to lose their power over time in the polymer particles. He couldn't know how the immune system would treat them. He would need to do long and complicated experiments in the lab to find out.

Langer and Folkman joined their big ideas in a common dream. It held risk as their path might have been the wrong one, and a lifetime might pass before they made a discovery, let alone developed a meaningful therapy. In pursuit of their dream to cure cancer, they took a first step to creating an enduring thing, and then they took the second step in committing deeply to struggle together.

In all this, Langer expressed his process in aesthetic forms that he, Brem, and Folkman found meaningful. These included equations that visually guided the mind along a simple path in a profound nonintuitive direction. Images that showed a microscopic life nobody imagined with stunning clarity. Limpid sentences that guided the mind straight from the reality everyone knew to the possibility of an altered reality. This new reality would be one that others aspired to but could never have imagined had they not read the sentences. These forms were not functional. They did not cure disease. At least not yet.

Folkman was able to place himself in Langer's situation and measure the courage of it. Folkman had a lab of medical interns and biologists who shared a medical culture that wasn't Langer's. Conversations, questions, and problems would arise, and Langer would be there with his head down scraping at calf bone or working on his polymer experiments. Folkman would visit his lab bench. He spent more time next to Langer than anyone else. Sometimes they flipped roles—Folkman liked to be the student, learning next to the teacher Langer. Seeing himself as the teacher gave Langer confidence and helped him press on during the most difficult first nine months. This was the period when Langer might have bolted and turned back to more familiar chemical engineering territory. Because Folkman joined him so fully on his patch of frontier, Langer dug in. Folkman seemed to have the same thrill when he saw the creator light go on in Langer's head that Langer had felt seeing the learning light go on in the minds of his high school students. Langer would work late, and Folkman would sometimes stick around to close down the lab with him. He told Langer to call him at home and vice versa, a habit Langer and Folkman would keep for over thirty years until Folkman's death. They drew deeply close as Langer struggled to isolate the angiogenesis inhibitor and prove his polymer hypothesis. He had no social life outside the lab.

One day Folkman walked a new student by Langer's bench. He nodded Langer's way and spoke loudly, for all the lab to hear. "Look at him if you want to know how to be great," he said. "Come in at all hours, work hard, scrape cartilage from bone, and have great ideas—it pays off."

Angiogenesis was easy for Langer to understand as a chemical engineer. In a way, he saw it more clearly than even Folkman and the other medical doctors. They knew far more about cancer than Langer did, but Langer knew far more about chemistry and making its principles actually work. Each stood in the other's gaps, which is the place where empathy creates surprising possibilities. As Folkman grew comfortable standing in Langer's gap, Folkman realized at last how revolutionary his angiogenesis idea was going to be.

Folkman had needed a maker in the lab. Discovery was important, but it took the mind of a brilliant engineer to actually create what matters.

Pioneers who arrive with fresh knowledge and an open aesthetic eye can be a godsend. That was what Langer turned out to be. He saw the tumor as a chemical factory that produced chemicals as cells do, and those chemicals diffused into surrounding tissue and reacted with other cells. Some of those could belong to blood capillaries. Reactions changed them and made them progress in the direction of the tumor, ultimately bringing nutrients to the tumor to help it grow. The hypothesis made evolutionary and chemical engineering sense, but because it was a new idea that lacked unequivocal scientific proof, the medical community treated it like superstition.

The angiogenesis inhibitor Folkman was looking for needed to be pulled out of an accessible place and isolated through standard chromatography experiments, in which you passed the material you'd cut out of cartilage through columns of specially designed gels. The gels slowed down the flow of the material. Some molecules traveled through faster than others based on their size. You knew the size of the molecule you were looking for and could therefore find it by seeing when it came out of the column. It could have been done before, but it took the new guy arriving in the lab to do the hard work to actually get it done.

"Folkman told me that inside baby calf cartilage I would find a molecule that inhibited angiogenesis." Langer didn't question Folkman and in a sense adopted the intuition of the medical doctor as his own. They shared an inductive hypothesis, while they agreed that if they could not prove it with deductive scientific rigor they would ultimately let it go. Langer worked hard for a year isolating the molecule that inhibited angiogenesis. Henry Brem jumped in to do the careful experiments it was going to take to prove that what Langer had isolated actually worked.

Brem had been working in Folkman's lab since obtaining his first degree in biology at New York University. He'd completed a year of graduate school work in biology at Harvard, then entered Harvard Medical School. He knew everything about the biology of angiogenesis that Langer couldn't read about in books or gather from his conversations with Folkman. Four years younger than Langer, Brem had raced through school. He became like a younger brother to Langer, absorbing Langer's maturity, benefiting from Langer's mentorship. Brem loved to grab lunch or hang out late with Langer, sharing what he knew, which served Langer as well. Brem was also open and "unknowing" enough to believe in the wild polymer hypothesis that made some others in the lab smirk.

Brem constructed and carried out all the in vivo experiments that would prove Langer's first paper, while

Langer gave Brem the insights and access that would later on help Brem develop a new brain cancer therapy.

"Pioneering brings you together in a rare way," said Langer. "I definitely would not have pursued my career without the bonds with others it formed, not just what I'd shared with Judah but with others in the lab, like Henry, and then throughout my career with my own students."

Langer seemed now to be doubling his risk at the frontier—digging for gold in two places that the world was convinced were void of treasure. He was on a journey that would take a year or two of living in the unknown to find out if the world was right.

Langer, Brem, and Folkman struggled to realize their dream, and when they thought they were upon it, they went public and expressed their discovery. They'd naturally followed the three steps of the Creator's Cycle: ideation, experimentation, and exhibition.

It is easy to see this cycle as purely declarative. We share what we have created and move on. For whatever reason, we believe others will be satisfied with what we have come up with. Creating in such a "productive" fashion belongs to more common ways of innovating, as in the development of a new cell phone or the writing of the latest installment in a successful book series. In cases like these, we can know what people want and can create it almost in the way of harvesting an apple, bringing it to market, and selling it. Pioneering creators can't do this. In the frontier research that belongs to the third way of creating, things that get exhibited are far from products.

These things—articles, designs, food forms—open up a public dialogue that sends creators back to their pioneering experimentation. Over time and iteration through the Creator's Cycle, the created thing evolves, and the creators evolve with it.

Two years after Langer arrived at Harvard Medical School, he published two seminal articles, one in *Nature*, the other in *Science*. The *Science* article showed the power of the calf extracts to arrest tumor growth—the first strong proof that angiogenesis was real and could be stopped. The *Nature* article, which he coauthored with Folkman, showed you could release large molecules from implanted particles made of polymers (ethylene vinyl acetate, which slowly released the statin) and therefore possibly stop cancer without regular injections.

Publication of the two articles led to the broader scientific conversation that Langer had hoped for. Other labs started trying to prove or disprove what he had shown with Brem and Folkman. Langer would actually submit nine research proposals to continue his work before he got the first one approved. The bond that held Langer, Brem, and Folkman together in the lab over the two years it took to prove their hypotheses grew stronger as they saw that their discovery had opened up new pioneering opportunities.

Langer's discovery led to the Gliadel wafer, a medical implant Langer developed with Henry Brem. Brem went on from Folkman's lab to become a respected medical professor at Johns Hopkins. Gliadel turned into a major breakthrough in brain tumor therapy. Folkman's angiogenesis hypothesis would meanwhile lead to a rethinking of cancer and, as of today, to eleven cancer therapies and 146 drugs in the clinic.

If working in Folkman's lab had made Bob Langer a discoverer, getting his ideas into cultural discourse helped eventually make him a creator of things that mattered.

Langer has published more *Science* and *Nature* articles than any living engineer. He has more patents (today around 1,300 issued and pending) and health care products on the market (over 100) than any engineer—ever. And yet, he says, his students matter most, more than any patent or invention. He has over three hundred students in academia alone. Every summer he invites them to his summer home in Falmouth on Cape Cod. Every other winter, many show up with their partners at a conference he organizes in Hawaii. In his lab, any student can see him anytime he is around. They dream, experiment, and share their results while he teaches them what he learned in the lab of Judah Folkman: to care in creating and to create to care.

Creating very new things tends to be a collective act. Many productive activities are solitary, of course, and success can sometimes come out of extremely anti-empathetic conditions. Frontier creating is different. Being long at a frontier with an anti-empathetic partner courts disaster. Teaming up at a frontier, surrounded by unpredictability and doubt, encourages an urgent kind of empathy that goes beyond the natural sensitivity to the point of view of the other, without which a colleague becomes a nightmare. It is closer to the empathy we have when we need to lean on one another to survive.

Creators maintain and grow this special aesthetic empathy through the Creator's Cycle of ideating, experimenting, and expressing. As with passion, empathy grows with each step, if through different actions. Step one, in shaping his dream he *built it with shared stakes.* Step two, while experimentally exploring his dream with others, he *bore with others the chance it would fail.* Step three, he expressed what mattered *with team participation.*

Where does aesthetic empathy come from?

Empathy originates in our brains by a combination of emotive and cognitive processes. Scientists long assumed that emotion and cognition tugged our brains in opposite directions. Emotions saved us from raging bears and devilish bandits. But to navigate the world and optimize our self-interest, we did best by keeping our emotions in check. Emotion was animalistic. Cognition was human. We thought, therefore we were. This simple vision of brain function lost favor among neuroscientists many years ago. Dividing up humanity into right- and left-brain contributors, the logical and the imaginative, the scientific and the artistic, happens to be based on an archaic misrepresentation of how our brains operate. Emotion and cognition work best in concert.

A remarkable system of the empathetic brain is the mirror neuron. Mirror neurons fire when animals perform an action, like grabbing a peanut, and they also fire when observing another perform the same action. They mirror in the brain of one animal what actually happens in the experience of another. Mirror neurons seem to play a role in imitation learning, perception of the intentions of others, and in the all-important emotion of empathy. More fundamentally, the neural origins of empathy include processes of cognition. A recent study by researchers at the Max Planck Institute in Germany, for instance, shows that a specific region of the cognitive brain involved in distinguishing perception of ourselves from that of others helps us delineate our emotional state from that of others and, when not required to process information too quickly, drives feelings of empathy. When this particular portion of the cerebral cortex functions improperly, people may fail to perceive or care about what others feel.

Aesthetic empathy may have little value when creating for short-term gain. To become the other is to have the time to discover who the other is, which can be a distraction in the nearest term. When we forget our own immediate personal interests, we can miss opportunities to win, profit, and advance a cause we are sure will benefit us. But when pioneering the future, immediate personal interests are themselves a distraction, and we more urgently need the eyes and ears and noses of those able and willing to be at the frontier with us (Figure 5).

Not every pioneering creator exhibits the passion and empathy of Ferran Adrià or Robert Langer. Some of the world's great creators, like the celebrated South African artist William Kentridge or the Nobel Prize–winning biologist Phil Sharp, are taciturn and circumspect. The emotive and cognitive dimensions that earmark creators as pioneers may often be not so obvious. But in the uncertain battle that determines survival of an idea, creative pioneers tend to excel in a way hardwired into the brain.

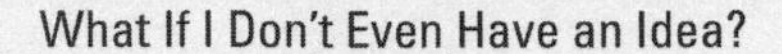

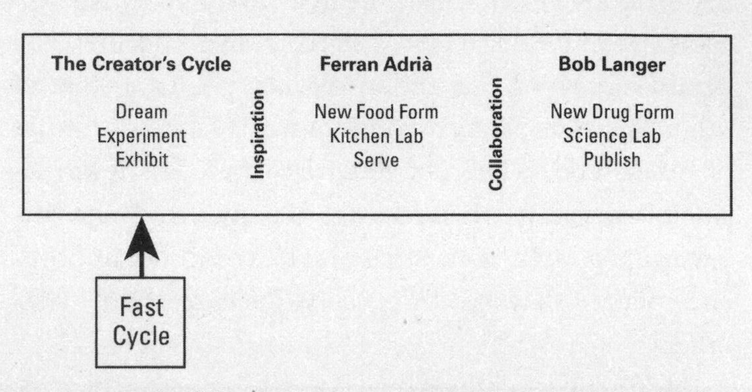

Figure 5. In the full course of creating things that matter, the first phase of creating is a relatively rapid iteration of the Creator's Cycle. Passionate curiosity generates inspiration and new ideas, and drives the hard work of getting others to buy into those ideas, while empathy helps creators listen well and, through collaboration, to adapt their ideas in ways that promote created things mattering more broadly.

In 2016, I served on the jury of the Rolex Awards for Enterprise with renowned marine photographer and explorer David Doubilet. The awards are granted each year to innovators who aim to better the planet and are among the first and most prestigious of the many so-called resource prizes that exist today. The night before the award ceremony, we began talking about ocean diving as frontier exploration. I asked him what qualified a diving partner. Doubilet put it this way: "They have to be nice. Goodness is the second most important quality of an explorer." I asked him what the first was. "Passionately curious. Good and passionately curious. You won't explore long without being both."

These thoughts echo throughout the history of innovations that endure. Albert Einstein described empathy as "patiently and sincerely seeing through the other person's eyes," while the African American novelist Zora Neale Hurston wrote of curiosity as "poking and probing with a purpose."

We do not take the risks to dive undersea or go to a knowledge frontier where we might discover the fundamental laws of physics or the poetry of fury without passionate commitment and frontier companionship. This truth tends to hold for all of us when we look to thrive in the radical newness of contemporary living.

CHAPTER 4

○

Experimentation: What Will I Do Now?

ONE YEAR, TWO STUDENTS IN MY HOW TO CREATE Things & Have Them Matter class came up with what seemed to be a good idea. During the semester, Tom Hadfield and Magnus Grimeland, both Europeans, had learned of a township in Zambia called Kantolomba on the outskirts of Ndola, Zambia's third-largest city. Kantolomba had fifteen thousand inhabitants and an outbreak of malaria infections. It also had, in the words of Hadfield, "no running water, no electricity. Not just no jobs, but also no money and no commerce. People lived in makeshift huts with roofs that collapsed every rain season. The town was embedded in between two huge graveyards, so funeral processions would pass through the dirt streets during the day." The students' idea was to import mosquito nets into the village, hand them out to five thousand families, demonstrate how to properly use the nets, and create a nonprofit to get the nets to all those who needed them. Over time they would encourage people to willingly make the nets a part of their homes, raise more resources, and scale the effort to help mitigate the spread of malaria. They prepared to partner with a Buddhist organization called Living Compassion, which was building a health clinic in the township.

Malaria is one of the deadliest infectious diseases on the planet. We actually know how to fight it and even how to eliminate the risk of it from an entire population, which is why many parts of the world are generally malaria free. Mosquito bites cause malaria, and the drug quinine, derived from the bark of the cinchona tree, effectively treats it. Poverty and lack of medical care are two of the obstacles that keep protection and prevention from many people who need it. As a consequence, around five hundred thousand people die each year from malaria. But another obstacle to protection is compliance, often related to a basic cultural awareness that malaria is a preventable disease. A recent study published in the journal *Lancet* showed that Cambodian villagers were not protected from malaria by a powerful insect repellant largely because they failed to use it.

My students applied for a travel grant from the Wyss Institute for Biologically Inspired Engineering at Harvard to put their idea to the test in Zambia over the summer. Their proposal had a breathless feel. Obviously, they were excited, possibly even overconfident in their ability to realize their idea, but there could be no question that they wanted to succeed, and they would learn from whatever new conditions they encountered. The grant came through (as did a grant from the Goldman Sachs

Foundation), and once the semester was over, they traveled to Kantolomba.

"On our first day," Hadfield explained to me recently by email—now ten years later an entrepreneur with a startup in Austin, Texas—as he flew with his wife back to Africa for a vacation, "Magnus and I arrived around eight o'clock. All the men congregated outside, gathered around 1980s ghetto blasters blaring bad music, while they downed their homemade moonshine. With so much alcohol being home-brewed, and the persistent heat, the whole place smelled like burnt chemicals all the time. As we drove in, every so often a man would dance or stumble in front of our car, put his hands on the hood, and stare at us." The feeling was, 'This might not be such a good idea after all.'"

Nothing in my class had prepared my students for what happened next. The class encouraged curiosity, an ability to listen, and—by frequent challenges to an expressed idea, frequent pivots from one direction to the next over the course of the semester—an ability to think on one's feet and move quickly in concert with others. It did not give students the time to research conditions that would meet them on the ground (my students might have taken a whole year of classes just to begin to understand the challenges and opportunities of their Zambian village). Knowing more than they did, my students might not have relied so heavily on intuition.

After a day of exploring Kantolomba, Hadfield and Grimeland decided they needed help to understand how their summer might be worthwhile to anyone but themselves. "Living Compassion organized regular community meetings that I attended." Hadfield explained: "At one, a woman explained that people didn't need nets, they needed food. 'We are hungry,' she said. 'The women will just sell your nets to feed their children, or the husbands will take them. And besides, what about the people who don't get the nets? You will divide our community.' So we tore up the plan and, based on what they said, decided to use our money to expand the food program that Living Compassion had started."

Able to admit their innocence, Hadfield and Grimeland used their intuition to address a problem as best they could (eventually, they decided to go door-to-door to do a census of kids, which helped them identify the kids in need of an expanded food program), and in a short time, they brought value to the villagers of Kantolomba.

Fast creation of things that matter to many is rare. It can happen when creators are faced with a crisis situation or upheaval and respond to it with fresh eyes, quickly creating something that lasts—as did Victor Hugo (during the social upheaval of French urbanization) when he wrote *Les Misérables*, Bill Gates and Paul Allen (in the early stages of the personal computer revolution) when they founded Microsoft, or Abraham Lincoln (at the height of the American Civil War) when he wrote the Gettysburg Address. In 1907, Pablo Picasso, who was twenty-six years old, painted *Les Demoiselles d'Avignon*, a rupture from everything that he, or anyone, had painted until then. Around the same time, Albert Einstein, also twenty-six, published four papers that changed the course of science, including his special theory of relativity. In a flash, Picasso and Einstein simultaneously expressed intuitions, grounded in the late nineteenth- and early twentieth-century technological reality of train travel and telecommunication, which gave new impressions of space and time mixing. Their creations expressed fresh conditions that were broadly shared and soon belonged to history.

Far more common is long sustained creation, as in the immense career output of Picasso, the careful crafting of the general theory of relativity that Einstein developed over the years that followed his first published article—or eventually the creative careers that my students began after they returned from Zambia. It is in this second and usually drawn-out phase of creating that other emotive and cognitive states matter even more than passion and empathy. As the next stories show, on the long trek between first ideation and ultimate realization of a dream, intuition, innocence, and humility help us when we discover that things are not as we had imagined, and if we are to change anything, our ideas need to change first.

Culture labs, which in this chapter range from schools and homes to science labs and startup companies, provide the protective support that keeps these critical aesthetic dimensions alive.

Reimagining the Digital—Aesthetic Intuition in Action

Danny Hillis created pinch-to-zoom, a gesture familiar to most of us through the daily tactile manipulation of our touchscreen phones. Few recent inventions have changed human experience in such a pervasive yet simple way. We place our fingers on a digital screen and stretch it out, zooming in on what we see. Billions use this created thing every day. And yet it did not come about with an intention to make a fortune or even to solve a particular problem. Actually, when it eventually caused an international technology storm, Hillis almost didn't even notice.

"I loved to invent, to take things apart and make something new," Hillis explained to me the first time we met at an airport in Singapore. We had each been invited by Microsoft cofounder Paul Allen on a vacation in Southeast Asia (in my case very indirectly, as I knew the director of Allen's new Frontiers Group, which supported pioneering biology), and we happened to walk out of the plane from Boston together. On introducing ourselves, we each used the word "inventor"—"Such a subversive title!" he laughed. We exchanged a few stories and eventually climbed into the car that took us to our hotel.

"Clocks especially interested me," he told me as we drove away from the airport. He has expressive blue eyes and a shaved head, and when he remembers the past, he shrugs his thick shoulders. "In the late 1990s, I proposed the idea of a clock that would last ten thousand years. Civilization has been around for about ten thousand years. It's likely we'll be around for another ten thousand. It seemed to me we should have a project that helped us keep that idea in mind, helped warn us from falling into short-sighted patterns. That's what led to the project Brian Eno (the British composer and Hillis's friend) later called the Clock of the Long Now. Hillis's project remained something of a myth in the minds of many for several years, discussed in technology, art, and design circles, while hardly imagined as an actual functioning clock. Then Jeff Bezos decided to help realize the dream of the Clock by building it on his property in Texas. According to public statements by Bezos, he has spent $43 million on the project.

Mathematician, computer scientist, and inventor, Hillis has an extremely intuitive mind. Unreal visions appear to Hillis at night, and he explores them during the day, as with pinch-to-zoom, which originated in his childhood dream of a giant map that portrayed the unusual places where his family came to live, a very unusual map that he could stretch out as far as he wanted and narrow in on things.

Intuition, or gut instinct, is one of the most critical dimensions to creating to matter. We build up intuition (which is anatomically linked to the caudate nucleus, one of the critical structures of the brain's reward system) over years of sensorial experience. Sensory data stream into our brains, even when we are not aware of it. Some data capture our attention, much do not. Our brains store these data with special recall access given to information—a face, an equation, a rut in the road—that has excited us. Emotive processes in this way guide our cognition toward what is actually useful and, in moments of split decision, may turn out to be critical. Emotions essentially tag sensory information that enters our brains, highlighting what matters and what does not. This tagged information has the privilege of being stored in our memory in forms less likely to be forgotten, and helps build up intuition (our experience-based models of what we might expect), which amounts to the very specialized expertise that allows us to "know the answer" without knowing why. Pioneers like Hillis learn to listen to intuition more than they listen to anything or anyone else.

Hillis grew up in Africa and India, with occasional stints in Baltimore. His dad was an epidemiologist and hepatitis expert at Johns Hopkins University, a career that kept his family moving to places like Rwanda, the Republic of the Congo, and Calcutta. Being itinerant put Hillis in the happy place of learning essential subjects with his biostatistician mother, and whatever else was to be learned by playing around with things that piqued his curiosity. He ended up at MIT in the mid-1970s when Cambridge was at the epicenter of the bit and chip revolution. After his undergraduate years, when he devoured the books of Jorge Luis Borges, Hillis pursued his master's degree while helping to develop the learning software Logo and then started his doctoral degree with Marvin Minsky, a pioneer in the field of artificial intelligence. Minsky had created the Artificial Intelligence Lab at MIT and had a bent for mischief and play that confirmed how Hillis thought about the creative process.

One of Minsky's inventions was a useless robot. You turned it on and a hand reached out of a box to flick the switch that turned itself off. Robots might do purposeless things, like us. Minsky also created very practical things, like the first artificial neural network, and with his popular book *Computation: Finite and Infinite Machines,* he showed Hillis how even sophisticated ideas, which took years of study to master, might find a broadly accessible and satisfying aesthetic form. "Danny became Marvin's protégé," says MIT architect Nicholas Negroponte, who later founded the MIT Media Lab. "Marvin eventually saw him as a son."

Meanwhile, it was in the Architecture Machine Group founded by Negroponte that Hillis began to think about, and play with, what Negroponte called the user interface. Negroponte recognized that computer interfaces were about to go far beyond textual interfaces. He imagined perfecting them for designing buildings and ended up fostering conditions that led to user interfaces being used to design just about everything else. Negroponte was passionate about the future of learning. With his MIT colleague Seymour Papert, the pioneer of constructionism as a project-based approach to childhood learning, Negroponte had taken over leadership of a new French government program to bring computer learning to the neediest on the planet. The ambitious if eventually ill-fated Le Centre Mondial in Paris included a high-profile program aimed to bring Logo language and Apple II computers to children in schools outside of Dakar in Senegal. Hillis got wind of the project, and in 1982, a year into his doctoral work with Minsky, he packed his things to go over to Paris for a couple months of tinkering.

Hillis imagined making a user interface adapted to kids who couldn't read. "Back in those days, we got paid in the Minsky lab to do one thing," Hillis explained to me during our second substantive conversation, which took place back in Cambridge at Café ArtScience. Darkness had fallen outside, and people streamed past the floor-to-ceiling restaurant windows, heading home or inside for a drink or dinner. "After doing what we were paid to do, we let our intuition run and invented things like email." He paused with a mischievous smile (Hillis and friends had intuitively played at a frontier, "discovering" things others would go on to translate into durable created things, or "invent"). "The stuff we got paid to do did okay. The intuitive stuff went on to change the world."

In Paris, Hillis developed the first prototype of the pinch-to-zoom screen. It took him a handful of weeks. Then, before Hillis headed back to MIT, Negroponte introduced him to Steve Jobs, who was in Paris to check on the status of the Apple II project in Senegal. Hillis and Jobs were each in their late twenties, Hillis the brilliant engineer, Jobs the brilliant businessman, both with enough intuition to imagine they might one day work together, compete against each other, and possibly even realize the other's dream faster than the original dreamer could. They spent the day together, ending with an adventurous car ride through Paris, with Jobs and Hillis sitting in the back of Negroponte's hatchback car, and Jobs jumping out at a series of stoplights to try to learn the phone number of a stunning taxi driver who trailed them. Jobs failed at this but succeeded in learning about something far more useful: Hillis's pinch-to-zoom idea.

It would take years for the seed of that idea to grow within the Apple universe, and by the time it did, it would turn into one of the biggest borrowings of invention in tech history (more on this later).

Intuition is everything at the frontier of human experience. The frontier might be any completely new circumstance—an adolescent showing up in a new school, an inventor proposing a new path to living on the planet, a climber of an uncharted mountain. The experience is new. It holds an opportunity and a risk. The pioneer needs to choose what to do without the advantage of precedent. Away from the frontier and in circumstances we feel we understand perfectly well (in an old and familiar school, polishing an invention that will only optimize an existing way of doing things, climbing a familiar mountain), intuition can seem much less useful. Our day-to-day behaviors and decisions can become rote. There's no question about risk to our advancement—let alone our survival. What is not logical can seem best to avoid. We do not plan a mountain climb with intuition. We calculate trajectories, analyze terrain, deduce supplies, and adopt a strategy that will get us to the top and back down. But once on the mountain, the climber, confronting a sheer face in an unexpected storm, with cold fingers, uses something else. Innocent to new circumstances, she pays attention and hopefully does the right thing. Her perceptions may be intuitive or deductive, and generally they are both. By intuition, she guesses a rock will hold her; by logical perception, she concludes it won't and reaches for a new rock.

Intuition and deduction, the wellsprings of art and science, are intertwined in the pioneering life.

A lifetime of adventurous experience had fostered Hillis's intuition. He now sharpened it even more. Pioneers become better by actually pioneering. Each new adventure reinforces what they have learned, in Hillis's case through the three steps of the Creator's Cycle. To facilitate *ideation*, Hillis placed himself in new conditions.

It was not obvious back in those days how people would want to interface with a computer. He wasn't completely sure how to make the interface. Paris was a new city. To understand his conditions and facilitate *experimentation*, he engaged in a collaborative dialogue with people more experienced than he was—talking with Seymour Papert and Nicholas Negroponte, among others, at Centre Mondial, and also Steve Jobs. Finally, he invented something new, a kind of new story that deepened his collaborative dialogue, *exhibiting* a first prototype of the pinch-to-zoom.

The steps were not new to Hillis. He had indeed taken the first two when he'd shown up at MIT and wandered into the Architecture Machine Group, surrounding himself with veteran pioneers. "There's nothing more important to me than being surrounded by people who are smarter than me," Hillis says without a sense of irony.

His Paris touch-screen map may have been the first pinch-to-zoom ever made. He might have patented pinch-to-zoom as a broad idea, and in a sense expressed his dream, and finished it. He believes that would have been a bad move. Universal ideas, in Hillis's view (such as the wheel, the computer, or wind instruments), belong in the public domain. They come to humanity in a moment and in a condition that human history has shaped; they are frequently expressed by many and are far more useful than any one person can possibly imagine. It is when you realize your idea in a particular way that matters to many others that you merit a patent. Hillis didn't yet understand his particular way. He was actually a couple decades ahead of his time.

Hillis came back to the hottest subject on the MIT campus: the neural network. Computers, it seemed obvious to Hillis, were ultimately to be wired like our brain to send, receive, and process information. In the brain, information goes from one neuron to the next in parallel paths that somehow link up to produce knowledge, emotion, and intuition, unlike what computers did. Hillis wanted to make a chip with this same neural capacity. People said that placing more than one processor on a chip would lower the computer chip's efficiency and prove to be a waste of his time. Hillis decided to test this theory. He started to make new kinds of chips, advanced from two to four to sixteen processors per chip, and by the time he'd proven that his parallel chip actually worked, he'd started a company with Marvin Minsky.

The day after the company office opened, a visitor showed up from California. "Richard Feynman reporting for duty," quipped the Caltech Nobel Prize winner, the first of many pioneers who would come to work with Hillis in those early years. Hillis's company, Thinking Machines, turned into a magnet for talent, and by the early 1990s, it had become the darling of the computer industry. Hillis had the smartest machine on the planet, and it was made by people who were themselves remarkably smart. Not only were Hillis's machines fast, they were also aesthetically striking. Cray computers, the established and principal competition for Hillis's new company, were unremarkable by comparison, squat and cylindrical, with parallel stripes that ran vertically from the floor suggesting their functionality. Thinking Machine computers were tall and sleek, looking like contemporary furniture. You wanted to know what was inside them, which made them popular in futuristic films (they showed up in the 1993 film *Jurassic Park* and in *Mission Impossible* in 1996). But already by 1993, the company was in free fall.

The intuition that had served Hillis in the frontier world of computer science proved, unfortunately, of less benefit at the frontier of free market competition. Thinking Machines had rushed to attain its market leadership by landing major contracts with the government agency known as DARPA (the Defense Advanced Research Projects Agency), responsible for far-out technological research for the military. Since at least World War II, MIT has had a strong record of supplying the US government with technology solutions, and Hillis and company benefited from this relationship. DARPA was known as the geekiest of government agencies, which made it an easy partner. But competitors like Cray saw the growing power of Thinking Machines as unfair noncompetitive access to the US government and began to lobby to break up the Thinking Machines monopoly at DARPA. Thinking Machines ignored the threat, didn't make an effort to tell its story to lawmakers, lost its government contracts, and, having failed to diversify, eventually went out of business.

The day his daughter was born, in 1994, Hillis declared bankruptcy.

In the long arc of creating, even the most prolific creators fail. The first failure can seem the worst. The best creators learn from failures and traverse them so fast they can seem to never actually fail.

Hillis was not yet in this mode. Depressed, feeling as if he'd failed his friends and colleagues, he started a consulting company and soon went to work for one of its clients, the Walt Disney Company. His friend, Bran Ferren, ran the Imagineers, the team of engineers and designers that had created the Disney Parks and the stories that would help Disney expand the park business every year from the time Hillis arrived until his departure with Ferren in 1999. Ferren opened a new position for Hillis called Disney Fellow and added the title vice president of R & D to give Hillis access to the business decisions that would make Disney, in the years under Michael Eisner, one of the most successful companies on the planet.

At MIT, Hillis had invented things he and his friends found personally interesting. He hadn't looked up and asked what mattered to others who were unlike him, nor figured out how to invite others into the adventure of discovery.

The Imagineers had ten useful storytelling rules. The first two were about paying attention to those around you:

1. Know your audience.
2. Wear your guest's shoes.

The next three insisted on clarity of purpose:

3. Tell a story that is organized and logically laid out.
4. Create a visual magnet.
5. Communicate with visual literacy.

And the last five argued for utter simplicity of message:

6. Avoid overload.
7. Tell one story at a time.
8. Avoid contradictions.
9. For every ounce of treatment, provide a ton of treat.
10. Keep it up.

Over the course of the 1990s, Eisner grew Disney Parks from its core Disneyland and Disney World destinations to parks in Paris, Tokyo, and Hong Kong, and expanded operations in Florida and California. The Imagineers were the creative minds behind the park expansion and content, and they invited Hillis to not just imagine new robots, which he did, but to listen in on the strategy behind the unfolding of the most powerful storytelling empire in history. Over the last years of the 1990s, Disney, which had made a historic deal with Pixar to produce three Pixar films (the first, *Toy Story*, had already rerouted film animation history), produced *A Bug's Life*, and put together the second *Toy Story* movie. As vice president of research, Hillis had access to decision making across the company, and he learned its storytelling rules.

Not only did Danny Hillis learn things, he also *made* things. On one occasion, his making went too far, and he learned a lesson that would later serve him. Hillis believed the expanding Disney Parks would benefit by having dinosaurs walking around park grounds. Where else could you possibly do this? He knew how to make them, and he believed that if he didn't actually make one of his wandering robots, the Disney leadership would never understand the opportunity. After working on his dream for many months, he invited Eisner and the senior leadership of the company into a warehouse on the Disney Studios campus. Everyone sat down before a big wooden box while Hillis described his vision for wandering robots in the park. It seemed obvious to everyone that the dinosaur robot was about to walk out of the box. Suddenly, the box behind Hillis collapsed. Eisner and the others watched, confused. Hillis had set up speakers around the warehouse. The sounds of powerful footsteps began to reverberate. They emanated from a big black void behind the fallen box. The sounds of footsteps grew louder, and all at once, a giant dinosaur stepped out of the dark behind Hillis. Eisner and the others fell back in their seats, then jumped up and scrambled away. Hillis had succeeded and failed. Clearly, dinosaurs in Disney Parks were going to terrify people. The project was killed. Hillis had effectively told his story using the ten Disney rules, only he'd lost track of the unspoken eleventh rule:

11. Tell a story that begs to be retold.

With his Disney Park dinosaur, Hillis had told a story whose powerful effect on those who experienced it was to ensure it would never happen again. It was the last time he made that mistake.

By 1999, Hillis decided to head back to the frontier.

With Bran Ferren, he started a consulting company called Applied Minds. Clients immediately lined up, and Hillis went into storytelling mode. He had completed step three in his learning of aesthetic intuition.

"Applied Minds is a place I started because I wasn't having enough fun at Disney," Hillis explained to an audience at the O'Reilly Emerging Technology Conference in San Diego, a few years after founding the company. It was March 2005, and *Craphound* blogger Cory Doctorow helpfully took notes. "I wanted to mix art, design tech, and science, and create version 1.0 of things." Hillis displayed a picture of a large warehouse, like the one from which his terrifying dinosaur had emerged. It held offices and electron microscopes and machine shops and computer-controlled cutters. He showed a NASA robot that moved in any direction and a snake robot. "You can connect a car, a robot, and they all work together. . . . Here's a cancer-simulator visualization that attempts to discover the chemical signature of which cancer drug works for which patient."

Then he turned to the idea he'd been looking to realize all of his life.

"I want a paper map that you can zoom in on by gesturing with your hands." He showed a video of what he'd invented. You could peel back layers and zoom in from satellite maps to street maps. The audience burst into applause. Suddenly, it made sense. The time had come to realize the pinch-to-zoom dream.

Hillis patented his idea and created with Ferren a company called Touch Table, which would develop the pinch-to-zoom technology for applications with Northrop

Grumman, among others, while Hillis moved on to other frontiers. There were many things to create, including the future of Google's search capabilities, which he actually alluded to in his 2005 conference in San Diego, with the quip, "But I can't really get into it because I'm out of time!"

Two years later in January 2007, Steve Jobs got up on the stage of San Francisco's Moscone Center to unveil the new iPhone touchscreen. Apple, said Jobs, had invented a pinch-to-zoom screen for its new iPhone. On the screen behind Jobs, the word PATENTED flashed. Jobs gleamed. He didn't mention Danny Hillis, nor did he speak of the brief trip he'd made to Paris just before he launched his Macintosh computer.

The effect was magical. The iPhone screen seemed magical too. Apple sold 1.3 million phones that year. Then a second smartphone appeared on the market seemingly with the same invention. The Samsung phone went on to sell very well, just like the iPhone did, and by 2011, between Apple and Samsung, roughly a billion people had a pinch-to-zoom screen.

Apple sued Samsung for patent infringement. Samsung fought back. While Hillis remained completely outside and mostly oblivious to the battle, the South Korean giant's lawyers pointed to Hillis's 2005 patent. "Hillis teaches distinguishing the number of contact points and determining whether the event object matches a gesture pattern," ruled the US Patent and Trademark Office in 2013. The Apple lawsuit was overturned. Appealed by Apple, the ruling was upheld in 2016.

The courts had decided. Hillis came first.

Hillis, who gained nothing monetarily from the Apple-Samsung battle (nor, he wrote me in an email, did he even try), makes light of the amazing story. "Inventions are like this. First ten thousand people have an idea. Then a thousand actually try to make it. A hundred get something that almost works. Ten get the idea to work. And one manages to bring it to the world. We call that one the inventor."

Alexander Rose has worked with Danny now for more than ten years. Based in Sausalito, California, across the bay from San Francisco, Rose has one of the most surprising jobs in America. As director of the Long Now Foundation, Rose leads the building and installation of the Clock of the Long Now, a massive machine coming together in California, soon to be installed inside a Texas mountain. The Clock, fruit of Danny Hillis's imagination, will be powered by a bellows on the top of the mountain, which will heat up in the day, filling with air, and shrink in the cold of night, blowing air back into the mountain where the temperature will remain constant. Breathing over the course of ten thousand years is supposed to keep the clock ticking. A series of bells will ring with the force of a suspended weight that visitors to the Clock of the Long Now, having made the trek up to and into the mountain, are supposed to wind into position during their visit. While the bellows system will keep the Clock functioning for a very long time, the mechanics of the chime system are contingent on human intervention. A full wind will apparently keep the bell ringing for around a hundred years, which gets to the possibility that people might forget about the Clock in times of crisis, war, ecological disaster, or whatever calamity the future brings.

Nobody exactly asked for the Clock of the Long Now. It does not solve a problem and will not obviously help the economy, and yet the Clock obviously matters to many (thousands of members of the Long Now Foundation send in checks from around the world every month to support long-now thinking). New things that matter last long after their creators walk away. They may not matter to every human being on the planet; inevitably, they do not. These things express something peculiar about human existence. The Clock of the Long Now has meaning to those who have a chance to wonder about the future. The novelist Michael Chabon wrote in a 2006 essay, "The Future Will Have to Wait," that "the point of the Clock of the Long Now is not to measure out the passage, into their unknown future, of the race of creatures that built it. The point of the Clock is to revive and restore the whole idea of the Future, to get us thinking about the Future again, to the degree if not in quite the way same way that we used to do, and to reintroduce the notion that we don't just bequeath the future—though we do, whether we think about it or not. We also, in the very broadest sense of the first-person plural pronoun, inherit it."

Long Now seminars in San Francisco attract large audiences most every week, like the one I attended with Rose, where the Danish economist Bjorn Lomborg spoke of his Environmental Assessment Institute's work to financially evaluate the costs and benefits of humanitarian efforts around the world. By any calculation, he admitted, the net present value of the Clock of the Long Now was zero. It obviously made no economic sense to create the Clock of the Long Now, even if one of the planet's smartest businessmen, Amazon founder Jeff Bezos, had backed the project.

Earlier in the day, Rose had driven me back to San Francisco from the metal workshop in San Rafael where the principal machinery for the Clock is being fabricated and tested. We were planning an exhibition of the Clock in Cambridge in the fall.

"What was it like to work with Hillis?" I asked Rose, an industrial designer, whose Long Now office sits on top of a popular San Francisco bar appropriately named the Interval, which is actually a kind of public culture lab of the Long Now Foundation and the site of Long Now seminars and exhibitions.

"Danny is incredibly smart, even brilliant, a word I don't lightly use," said Rose. "But what most people don't realize is that he can be as comfortable chatting with a machine shop worker in Texas as he is discussing arcane mathematics with the winner of the Fields Medal. He gets who he is talking to and adjusts what he says in a way that makes you sense he is always learning. Smart as hell, he is intuitive beyond belief."

Reimagining Social Behavior—
Aesthetic Innocence and Humility in Action

Richard Garriott created the first massive multiplayer online role-playing game, *Ultima Online*, and helped build one of the largest entertainment markets today on earth.

With Garriott's creation, the era of massive online social networking began. Virtual real estate started to sell for actual money, and real-life social behavior started its extraordinary virtual entanglement. Garriott pioneers the social gaming practices that are changing how we think and learn. A pioneer of the virtual, he is also an explorer of remote earth and space and an avid proponent of the benefits of the personal avatar, a notion he happened to create on his path to making the first games with moral rules of play.

Garriott lives with his wife, five-year-old daughter, and three-year-old son in a New York City townhouse that is like a shrine to his creative life. He walks visitors through it with the spry gait of a man who discovers things for the first time. On the ground floor he has arranged artifacts that progress from the first formation of our solar system through early humans to the future, including the space suit he wore when he shot off into space in 2008 (an adventure Garriott undertook while his personal net worth fell from tens of millions of dollars to nearly nothing with the stock market crash). On the next floor he has exhibitions dedicated to his passions for magic, haunted houses, medieval mores, and space travel, and on the top floor, in his compact loft office, he has relics saved up from his entire creative output, including his twenty-eight early video games, science fair projects he won as a child, and an Apple II computer that still plays his original *Ultima* games.

Garriott's house, like his life, exhibits the fourth dimension of creating to matter: the innocence that allows us to step outside the routine of life and discover.

Richard Garriott grew up in a Houston suburb of NASA scientists. When he was twelve, his dad, Owen, an early scientist-astronaut, entered orbit. Owen returned to orbit in the first Spacelab when Garriott was twenty-two. By then, Garriott was already a millionaire, and Buzz Aldrin, the moon-walking astronaut, came asking for investment to start a new venture. Garriott invested in Aldrin's company along with other "astronaut start-ups," lost most of it, and learned a lesson about the pioneering life. It didn't matter how smart you were or that you had pioneered before—every new discovery takes preparation. Garriott became a ferocious autodidact.

As a kid, Garriott didn't care much for formal learning, just as Ferran Adrià didn't care about food. He had a passion for playing *Dungeon & Dragons*, loved to compete in science competitions with the help of his mother, and to get through school, developed a technique of writing crib notes in the runic script he'd learned from reading Tolkien's *Lord of the Rings*. The trick helped him squeak by with passing grades, until he got a break.

When he was fourteen, Garriott moved to Palo Alto. His dad had taught at Stanford before coming to Houston. While he would remain a NASA scientist until the end of his career, Owen had decided to take a year-long sabbatical to brush up on his physics. As a consequence, Garriott spent his freshman year in a high school in Palo Alto with four fully functioning teletype computer terminals. When he returned to his hometown of Houston, there was only one such computer, and nobody actually knew how it worked. Computers may have been central to NASA success, but in those days, NASA kids had no chance to learn about them. For Garriott, this now changed, and he couldn't learn fast enough. By the next year, after he returned to Houston, he had a competency that made him without peer in his high school. His teachers encouraged his newfound skill, telling Garriott that if he wanted to use the school's computer to create, document, and demonstrate to them the kinds of novel fantasy games he loved to play, Garriott might count it toward his foreign language requirement.

Garriott started creating code, and soon had the attention of friends and family. By 1977, he had a cadre of admirers. Among them was the manager of the local Computerworld store where Garriott worked evenings after school. The Apple II had just appeared, without much content, and Garriott's boss said that if his geeky employee wanted to make a video game for Apple II, the store might be able to sell a few. Garriott took a break from his school's mainframe computer, for which he'd already written two dozen video games. It took him about six weeks to make his first Apple II video game.

He called it *Akalabeth*. The idea was to kill underworld creatures. *Akalabeth* went on sale in 1979. Within weeks, copies had started to sell in Computerworld stores around Texas on discs inside ziplock bags. The gaming company California Pacific eventually bought the *Akalabeth* rights to sell the game nationally, and by the time Garriott had turned nineteen, thirty thousand copies of *Akalabeth* had sold around America. Garriott, who made five dollars on the sale of each game, was by then doubling his father's salary. It seemed like a miracle. His parents encouraged him to try his luck as long as it lasted. So

Garriott worked away in his bedroom on a game for California Pacific, which he called *Ultima*. It was a huge hit. Two years later, he published a second version. By then, he was making hundreds of thousands of dollars each year while paying for his college education at the University of Texas at Austin. He was also learning the bad business practices of the early gaming industry. After twice having his publishers stop paying royalties to better afford recreational drugs, Richard and his brother, Robert, who had a degree in business management from MIT's Sloan Business School, decided to start their own company, Origin.

Origin launched *Ultima III* in 1983. The game sold well. And for the first time, Garriott started to receive fan mail. Reading his fans' feedback changed everything.

As a kid it had never occurred to Garriott that his desire to create virtual worlds would be the path to making a lot of money. Even when the money began to flow, he'd worked away as before, engaged in a cycle of imagining virtual worlds, developing them with friends and colleagues, and exhibiting them to the world in the form of games people bought and (for reasons he didn't yet fully understand) played again and again. The successful sale of his games had until then confirmed for him that what mattered to Garriott mattered to many others. But even a cursory read of his fan mail showed Garriott that his creations mattered to others for a reason he hadn't suspected. The goal of *Ultima* was to kill the bad guy. The longer you played, the more powers you assumed and the better your chance of winning. All video games used that basic principle. Good triumphed over evil. But in reading his fan mail, he discovered that good actually meant nothing to his fans. They actually liked to kill the monsters *and* the villagers in his games. Winning was all that mattered to them. They didn't distinguish between good and bad deeds.

Garriott's virtual world seemed to him now false. If people lived in real life as they did in his games, the real world would be utter chaos. He wanted to fix the problem and wasn't sure how. He had a weak grasp of the history of Western civilization, knew about the Ten Commandments, and had a vague understanding of natural human rights. He didn't know how to make a world all by yourself and imbue it with moral values. Garriott was innocent. His childhood hero Buzz Aldrin had been recklessly innocent when he'd asked Garriott to invest in a startup company he'd imagined, and as a consequence Garriott had lost his money. However, innocence in Garriott's case had value. It motivated him to learn.

Aesthetic innocence is how children, artists, scientists, and pioneers of all kinds bring to the world fresh insights and revelations—the kind of free-spirited unknowingness that promotes intuition. Insensitive or mindless unawareness is different. The risk of it is why we teach away from another kind of innocence in our best schools and warn against it in our most powerful corporations. Being recklessly innocent, we fall into traps. We say and do things we later regret, experience things we have no idea how to express. We seem to exist in a bubble of perception that threatens to pop. Aesthetic innocence is different. It combines with passionate curiosity, empathy, and honed intuition to promote wonder, inspiring us to pause before the unknowable and look more deeply at it than we otherwise would. We may be highly educated, or deeply experienced in other circumstances, but we ignore all this, and manage to pay attention to what we are seeing, as if for the first time.

The value of innocence in the collaborative experience of creation gets to the iceberg analogy of brain function. The tip of the iceberg is like the conscious brain, while below the water is the principal mass of the iceberg—the unconscious brain, the repository of deep-seated memory, motor function, and intuition. Conscious thought, feelings, and memory represent a tiny fraction of brain function. Imagination emerges at the waterline—a free or "innocent" exchange of information between the conscious and the unconscious brain. This is valuable as most of what our brains do happens underwater. From unconscious brain activity comes the intuitive knowledge—the result of running the present data through the predictive models we have built up over a lifetime—that our brains draw on when cognition comes up with a blank. It is like the experienced and well-prepared mountain climber who suddenly finds herself on a remote ledge at the start of a winter storm. Weather conditions and her chance position place her in a situation for which no amount of preparation could have prepared her. She must quickly decide whether to move to a higher altitude camp, which is close to her though accessible by a path she has never taken, or a lower altitude camp, farther away though accessible via a now familiar path. Cognitively innocent to her conditions (while emotively educated by years of climbing experience), she doesn't hesitate to ignore the information she has learned in her weeks of preparation and to trust the rich bank of information her brain has accumulated as an experienced climber.

As Matthew Lieberman points out in his book *Social*, most of our brain mass grows after birth, while we are in the incubative social context of family and friends. So significant is the human's postbirth brain mass growth that, at least in proportion to our body mass, our brains get to be far larger than those of any other animals. On a sheer mass basis, our brains remain on the small side. Elephants have bigger brains than we do. Mature whales have the biggest brains of all. That's because the brain controls bodily function. Big body, big brain. It is when we divide the mass of the animal brain by the mass of the animal body that we get a number reflective of relative brain potential, the ratio of brain to body mass. And it is on this scale that human brains come out on top. Our closest competitor appears to be the bottlenose dolphin, at about half the human brain to body mass ratio. Most of our excess brain mass can be attributed to the neocortex, the gray matter that covers the surface of the forebrain, possessing several distinct anatomical regions. One of these, the prefrontal cortex, is essential for cognition, and another, the posterior association cortex, is where auditory, visual, and somatosensory association meet. We process sensory information and ultimately perceive or interpret the world around us—aided by our sophisticated cause-and-effect models of the origins of stimuli we can detect—better than other animals do.

When we recognize that we are at risk in our innocence, we pay better attention to this information. Those who create over long periods of time and derive pleasure from the process, like Garriott, tend to value innocence for this reason and try to maintain it throughout their lives. They develop it through the same process by which they develop the other dimensions of the pioneering creative mind. First, creators *make things they want with few expectations* of how valuable these things might turn out to be. What they will do next is entirely dependent on what they discover from these new things they have created. Second, they are *stubbornly attentive and learn with fresh eyes.* Finally, they *express what they discover, thrillingly,* as if nobody has made anything like what they have just made.

People often grow up in settings that protect and promote innocence, as in a family that encourages storytelling, dreaming, and conversation or performance. Later on, once grown up, they may still listen to stories, read novels, or watch a ballet performance, but their ability to fully lose themselves in these adventures can be weighed down by personal responsibilities, professional challenges, and financial worries. Creators of things that matter manage to bring back childlike innocence through the creative process itself.

Garriott had taken his first step in honing his innocence when he'd noticed the amorality of game play. It was the summer of 1983. He began to read the works of Rousseau and Voltaire, of Kant and Nietzsche, of Sartre and Wittgenstein and Foucault. He studied Eastern and Western religions and the perspectives of Freud and

Darwin. The essence of morality was reduced in his mind to three values, captured best, he felt, by the writings of Kahlil Gibran: truth, compassion, and courage. He decided every moral value could be built from these three. Garriott dreamed of a game that motivated players not to kill, but to live valorous lives. It was a challenge he gave himself. His brother, Robert, called it a bad idea. Garriott was already selling games. People liked what he did. He and his brother had bet everything they had on the hope that their products would sell. Creating a whole new motivation for video game play seemed like a useless risk.

Garriott disagreed. He wanted a game that mattered. He hoped to draw a line between moral and amoral game play, and he suspected others would too, even if they didn't know it. Above all, he wanted players to take their virtual lives as seriously as they did their real lives. He at least did, and it maddened him that so many others, who were to him like family, did not. Unchecked amoral behavior was unacceptable to his players in their real lives, so why did they embrace it in their game lives? He suspected a flaw in game design and wanted to fix it.

Garriott had meanwhile come across the Hindu notion of "avatar," a word ascribed to the person who showed up on earth as the embodiment of a god, like Krishna. It occurred to him that his players should see their own game characters as a personal emissary—the game character should be an avatar! Thus the new *Ultima* game acquired its name, *Quest of the Avatar. Ultima IV* invited people into game play, fully, with all their hopes and beliefs, and the word "avatar" took on its contemporary meaning.

Ultima IV proved a smashing success. It confirmed Garriott's intuition and rewarded his innocence. The first role-playing game that departed from the hack and slash dungeon style of play, *Ultima IV* would be called the second-best video game of all time a decade later by *Computer Gaming World*.

Garriott had made what mattered to him, in an aesthetic form that mattered to others.

Ultima IV reached the market in 1985, the year after Steve Jobs announced the launch of the Macintosh. Garriott had made all his games so far on the Apple II platform. By 1985, Apple II sales were in continuous decline. Macintosh was the future, and Garriott had created one of the best video games ever in the wrong format.

He needed now to code his games for the personal computer, or his growing empire would be doomed. Although at this point in time it seemed doomed anyway.

By then, large gaming companies, like Electronic Arts, were starting to dominate the retail market for games. The rocket growth of the new industry had produced a myriad of small companies. Garriott and Origin had entered the era of consolidation. It became ever more difficult to remain independent. Origin produced two more *Ultima* games, *V* and *VI*, the first on Apple II, and the second, which came out in 1990, for PC. Making the transition to PC saved Origin long enough to allow Garriott and his brother to sell the company in 1992 to Electronic Arts for $30 million. The entire Origin team went to work creating for Electronic Arts, employee numbers swelled, and then they collapsed.

Ultima VIII came out without proper preparation and failed.

What had happened to Danny Hillis in the demise of Thinking Machines now happened to Garriott. His creations did not matter as much as he had hoped. After the sale to Electronic Arts, Garriott, while considerably richer, was a disgruntled employee. Everything he did, or his team did (because they had the overhead of a larger company), carried a larger price tag—too large. The cost of experimentation went up. He had to fire people.

On one occasion an employee who had worked for years with Garriott placed himself on the floor while a colleague outlined his body on the floor with tape. Over the next days others placed flowers on the outline.

For Garriott, the situation was beyond sad. It seemed actually tragic and might have ended a brilliant creator's career. Instead, as proved true for Hillis, defeat became the pretext for Garriott's most transformative creation and success.

Aesthetic humility, the fifth dimension to creating what matters, is the recognition that to discover at a frontier is to ever justify to others the worth of being there. Creating what will be sustained and contribute to the sustainability of other things is not simply a question of a special passion, but of a contract with others who agree to support what is highly personal, experimental, and risky. In the course of the aesthetic cycle, we recognize that we do not matter alone. We need others and look up. Alone we are as vulnerable as our created thing. It is not just that our creation might die without the interest of others; it is that we might die with it.

Humility in creating relates to the effective function of the social human brain. Scientists long assumed our special brain size evolved to better carry out logical function. In the 1990s, the evolutionary anthropologist Robin Dunbar decided to test the hypothesis. Dunbar explored three explanations for varied brain to body mass ratio across animal species. One was individual innovative potential, or cognitive ability. Another was social learning capacity, related to the large limbic lobe. The third had nothing directly to do with brain function, but tied to a specific feature of our environment, our social network size. Of the three, social network size correlated best across all animals. Mice, bears, dogs, and cats all tend to live solitary lives. Bottlenose dolphins swim in pods of up to twelve dolphins. The human social network size is around 150. This is also a typical size of the small enterprises that account for around half of US gross domestic product. Our brains have grown to their large size, at least in part, to benefit from a social group. Naturally, we care about the group, quite a lot (humility and empathy are like the flip sides of a socially engaged creative life).

If we touch a hot iron, we feel pain. Scientists use fMRI to map the pain we feel to particular regions of our brains, including the thalamus, the hypothalamus, and the prefrontal cortex. There is an electronic game called *Cyberball* where a player is asked to toss a ball back and forth with two other artificial-intelligence players. After a while of throwing the ball back and forth, the two programmed players suddenly keep the ball to themselves. The real player feels excluded. fMRI imaging maps the social pain felt by the excluded player to precisely the same regions of the brain that mediate the pain felt by touching the hot iron. Social pain literally hurts.

The notion of an extended social body goes even further. fMRI studies of the brains of people who witness a mate suffer pain show again the neural signatures of physical pain, as if the pain were their own. And when sufferers of social pain take a painkiller, like Tylenol, they can feel less pain.

Losing an intimate connection to a friend is not as painful as losing a limb, nor does seeing my friend cut his finger hurt me as much as when I cut my own. We do not all feel distributed pain to the same degree, or in the same way all the time. Mothers, neuroscience research has shown, will generally feel the pain of their infants more powerfully than, for instance, a football coach will feel the pain of an injured wide receiver on his team. But in general, we do seem to feel pain that is outside our own bodies to the degree it relates to the disruption of the harmony of our social network.

We feel social pain in the way we feel social reward. Recent neuroscience research has shown that our brain's reward system fires when we win money at a lottery game, but it also fires when, having won, we see someone else win, even though we have now lost. The human brain reward system embraces the idea of fairness. We like to do well for ourselves, but we do not enjoy the reward as much when everyone around us does poorly. Personal pain and gain is not all there is; we truly do care about others. Seeing need around us, we tend to create what matters not just to ourselves, but to others as well.

Humility inclines us to adapt our created thing to heighten the probability of benefit to our clan and is learned in the course of the Creator's Cycle. By sensitivity to failure (in step one), willingness to adapt the created thing (in step two), and continual editing (in step three) we simultaneously increase the value of what we create and perceive social reward.

As the Internet grew, Garriott dreamed of creating a kind of massive multiplayer online role-playing game, or what he called the MMORPG. He raised the idea with the management of Electronic Arts and began fighting for the chance to create it from the moment his *Ultima VIII* flopped. How to rationalize it? No obvious business model existed. Multiplayer video games, the so-called Multi-User Dungeon or MUD genre, had existed since the 1970s. The best MUDs had 15,000 subscribers, far below what Electronic Arts needed to justify investment. Having lost credibility in the failure of *Ultima VIII*, Garriott managed to convince no one. Intuition and innocence hardly served him. Nobody cared about this frontier, and particularly nobody wanted Garriott to spend his Electronic Arts revenue exploring out there. He and partner Starr Long lobbied Electronic Arts three times over a year to back *Ultima Online*, and was each time told "no."

In 1995, Amazon started to catch attention with what seemed to be a parallel business hypothesis. If books could sell online, Garriott reasoned, games might too.

He finally convinced management at Electronic Arts to let him overspend his development budget by $250,000 to make a prototype and with it somehow prove that a market existed beyond the 15,000 users of a successful MUD game.

With a ragtag team, he went to work on a beta version of *Ultima Online*. He couldn't afford testers, so he came up with the idea to ask *Ultima* fans to send in $5 to get a CD with the first beta. They put up EA's first ever website to make the offer, and 50,000 fans sent back a check. Electronic Arts got the point. The parent company now poured resources into the project while Garriott and his team doubled down. The beta testing began. It went on for a year, while the *Ultima Online* team addressed countless bugs.

The beta revealed fascinating new online behavior, ranging from collective online player protests to deep interpersonal bonding. On the last day of the beta test, Origin announced that servers would be wiped clean at midnight. *Ultima Online* would go live the next day. Garriott's avatar, a heroic unbeatable character named Lord British, traveled that last day around his virtual world to say farewell to all the beta players. And then, five minutes before the end of the beta, the inconceivable happened.

Lord British stood on a bridge before Castle Blackthorne addressing a group of beta users. A character named Rainz threw up a fireball. Lord British, fearless, stepped into it. Garriott's avatar was supposed to be immortal, but in the rush to get the beta version finished,

Garriott had forgotten to tag Lord British with immortal powers following the latest reboot.

Lord British died.

Pandemonium broke out. The beta of *Ultima Online* ended in a massive slaughter, sometimes referred to in the gaming industry as the most memorable event in the history of MMORPG play.

The success of the *Ultima Online* beta, and the death of Lord British, announced that virtual world societies were going to be as popular as they would be unpredictable. In a way that nobody could have guessed at the time, massive online social interactions began to change how human civilization worked. Surprising social dynamics started to happen, as hundreds of thousands of users began to play *Ultima Online* in 1997. Eventually, these dynamics became the hallmarks of online game play for tens of millions of people around the world.

You advanced in *Ultima Online* by purchasing land and swords. Your ability to purchase these things required you to acquire gold wealth as you played. Some players, however, didn't have the patience. They wanted to own things and advance in the game to have more meaningful experiences without investing the time to earn them. Those who did invest the time to build gold wealth discovered they could sell virtual gold for real money on sites like eBay, and thus a new economy of virtual gold exchange came into existence. Companies emerged in China with employees who spent all their time playing *Ultima Online* to acquire the gold they could then sell for real currency. Virtual gold mining companies created artificial intelligences to play *Ultima Online.* AI players started to kill real players to better hoard riches. Drug dealers started to use *Ultima Online* to launder money. Those players who entered the virtual world to play by the rules created tribes and societies to protect themselves from AI players and other villainous threats, and within these societies deep bonds formed, marriages took place, and before long the virtual world of *Ultima Online* started to resemble the real world from which it had emerged.

Seven years after the launch of *Ultima Online*, *World of Warcraft* appeared. Around the same time came *Second Life.* The gaming industry grew over the next two decades to nearly $100 billion in 2016, three times larger than the entire film industry.

Game play took on radically fresh meaning. Not only did it represent now the largest entertainment industry on the planet, it became an enduring basis for social networking and, remarkably, a platform for twenty-first-century learning. Multiuser online gaming (while it could promote gaming addictions and other neurological dysfunctions), when designed for learning came to promote curiosity, relieve stress, allow individualized learning, facilitate technical skills, and permit literary-like engagement across ethnically and morally challenging lines. The creator who had tricked his teachers with runic script to get through school had at last helped create a platform for learning that got around teachers altogether and placed every learner back in primitive

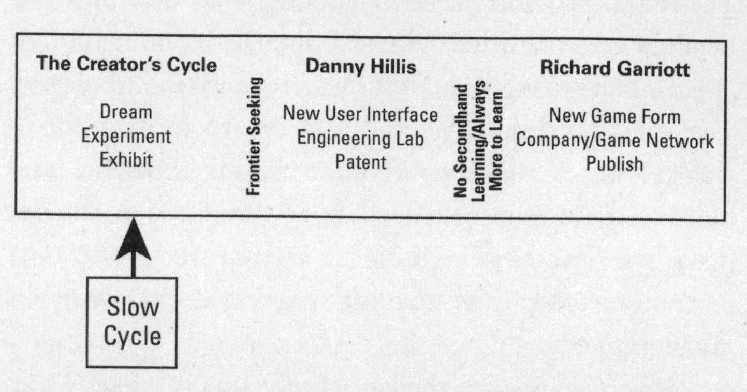

Figure 6. The second phase of creating things that matter tends to be the longest and most challenging. This iteration of the Creator's Cycle slows down. Durability and survival in frontier conditions comes through intuition—a kind of gut intelligence—and willful innocence. This keeps pioneering creators, who are deeply aware of the risks they take, attentive and avid learners. Humility helps them learn fast, adapt, and not rely on a confidence that belongs to more familiar circumstances.

learning conditions of human curiosity and survival where aesthetic dimensions get learned (Figure 6).

I walked out of Richard Garriott's Manhattan town house after my last conversation with the gaming pioneer, remembering it as a shrine to a long, bumpy, and giddily varied creative journey guided by the intuition of moments that are past, the innocence of moments to come, and the humility to change paths on a dime as the human experiment moves on.

Mass Improvisation Culture

Hillis and Garriott create in the aesthetic way of Adrià and Langer. Their work exemplifies the second phase of creating, the long stretch between ideation and exhibition or experimentation—where many creators lose heart. These creators of digital things that matter take their time to figure it out, hanging around at new frontiers, running from secondhand experience, and endlessly learning. They exercise and reinforce all three dimensions of aesthetic creating that help them endure—intuition, innocence, and humility—and eventually discover.

They are improvisers, like all of us. Indeed, among the many consequences of digital creation, improvisation or behind-the-scenes experimentation is increasingly the crux of the performance. The traditional line between creator and audience vanishes, as it did in the death of Lord British.

Traditional cultural forms that have prescribed conventions—such as classical ballet, opera, and stage theater—have lost audience numbers, while fresh, raw, unscripted cultural forms with active audience participation—like reality television, improv rap, and online social networks—have grown exponentially. Mass audiences today are not just willing to enter into a creative conversation with inventors of things that have never existed before, they often take over the conversation. User-generated content has exploded as a consequence.

Theater may be dead, as the theater director Paul Sills liked to say, while theatrical improvisation seems to increasingly characterize what is culturally alive. Sills was the first director of Chicago's Second City, the famous improvisational theater group whose actors went on to create some of the most resonant contemporary comic media programs from *Saturday Night Live* to *The Colbert Report*. His mother, Viola Spolin, founded American improvisational theater in the first half of the twentieth century. "We learn through experience, and by experiencing," wrote Spolin in her classic text *Improvisation for the Theater*, which she finally published in its definitive form after the start of the Second City, "and no one teaches us anything."

In Spolin's words, experience is an environment penetrated by intellectual, physical, and intuitive sensitivity. Of the three, intuition is, for Spolin, the most essential. With improvisation, idea development resembles playing a game in which the idea, or the point of concentration, is a kind of ball.

We play ball. There is no star, no separation between one player and the next. The audience is as much a player as the player the audience. In effective improv, passion, empathy, intuition, innocence, and humility all reign, and improv—aside from being near the heart of contemporary culture—becomes a metaphor for the collaborative human experience of creating to matter.

CHAPTER 5

Exhibition: Minding the Ball

Irene Nicolae, a student in my How to Create Things & Have Them Matter class, had an idea to create a drink that somehow helped your immune system. At first glance, Irene's idea seemed hardly new. It was ground zero for hundreds of new commercial products in the growing category called functional foods, and it reflected an unending stream of home remedies, from spices like vanilla and turmeric, to food fads like the Paleolithic diet.

However, Irene happened to be studying molecular and cellular biology, and she'd learned about what goes on in our stomachs that until recently nobody had known. Just five years before, the Human Microbiome Project had published the first results of the genetic sequencing of the human microbiota. Bacteria, fungi, viruses, and other multicellular organisms live inside our bodies, in our mouths, in our gut, on our skin, in the nose, and other places. Owing to this microbiome, a surprisingly decisive realm of our immune system living outside the walls or epithelia of our organs, we are made up of many more nonhuman cells than human cells from birth.

Recent research, Irene had noted, suggested that the use of antibiotics and various chemical additives in modern foods harmed and diminished the human gut microbiome, leading to increased rates of obesity, asthma, allergies, and autoimmune disorders. Scientific studies had also shown that by introducing probiotic strains of bacteria into foods, as Dannon had done with its Activia yogurt, new food products might change the genetic expression of resident bacteria related to metabolism. But scientists had also noted that the survival rates of live bacteria in market preparations of probiotic foods was low, far lower than what is required to have a positive effect.

Irene saw an unmet health care need, and an opportunity. She wanted to make a drink that would be delivered fresh, keeping alive the billions of bacteria needed for good human health, in particular those bacteria that had been shown to be separately implicated in obesity, Crohn's disease, and basic immune disorders. Should young people begin to eat fresh natural foods and supplement these with fresh engineered foods containing sufficient numbers of probiotic bacteria, human microbiota might be restored and curb the upward spiral of immune-related diseases. Irene had read enough of the research to convince her that if she could keep the bacteria alive in her microbiome smoothie, and get people to actually consume it every day, she could start a successful company that might benefit human health.

By the middle of the semester, she had ordered bacteria online and started working with the learning labs in the Harvard School of Engineering and Applied Sciences to grow and prepare them into smoothie food forms. She was planning to test different formulations, with alternative kinds of fruit, natural sweeteners, and dairy products like ice cream and milk. She would then validate their survival in the Learning Labs, before asking friends to try them out. As a precaution, she asked some of her friends what they thought of her idea. A couple actually drank smoothies each week. Most were intrigued, but none seemed likely to change how they ate, or give up drinking the smoothies they loved. Irene's idea was cool, and maybe one day it would help people live more healthy lives, but her friends were all pretty happy with the things they ate.

Irene seemed to have a good idea. Conceivably it could improve health care without invoking doctors and nurses. The science made sense. It might even lead to a very good business. But nothing would happen if people didn't really love the taste and overall experience of her smoothie (getting people excited about drinking gut bacteria was a particular challenge). The stability of her bacteria seemed secondary by comparison. She needed to give her friends something they wanted.

How would she learn to do that?

Late in the semester, I invited Irene and her

classmate, Shirley, whose idea also related to the aesthetics of changing consumer behavior, to go see an unusual theater production in Harvard Square called *The Donkey Show.* We were accompanied by the executive producer of the American Repertory Theater, Diane Borger. Borger had come to the A.R.T. from London with the production *Sleep No More*, the brainchild of the British theater group Punchdrunk. *Sleep No More* took place in an abandoned warehouse, an improv production with a storyline from Shakespeare's *Macbeth*. It had gone on to Manhattan, enjoying a success that *The Donkey Show*, which had been running between New York City, Europe, and Boston for nearly twenty years, had helped pave. *The Donkey Show* reframed the story of Shakespeare's *A Midsummer Night's Dream* into a 1970s disco experience, combining songs people loved with choreography that blew away the wall between actors and audience. The production did something Irene wanted to do: it revealed the beauty of Shakespeare's work to a contemporary audience that might not otherwise have perceived it. *The Donkey Show* had broken theater attendance records and changed public behavior by exhibiting Shakespeare's work in a resonant contemporary language. If this could be done with an Elizabethan play, why not with a microbiome smoothie?

On our arrival, Borger sprinted us through the dense crowd. The music pounded. "See if you can tell when the play begins," she yelled over her shoulder. I assumed that would be easy. We sat at one of the few tables that overlooked the crowd. Our view was perfect. Nothing escaped my attention, I thought. Fairies hung from banisters and leapt over railings. The movement of scantily clad muscular men seemed to be building up to something. I noticed the fairy queen. Actually, she had been here for a while—I'd missed the start of the show. Midway through the production I headed to the bathroom, crossing paths with the fairy queen, who had changed into another costume. The show was everywhere and nowhere. The rest of the performance flew by, with Irene and Shirley dancing at one point with the fairy king, Oberon.

Getting people to do the unexpected, and to enjoy it, changing how people live, means paying attention to what others care about and giving them what they never imagined. It is a kind of ball game creators play, tossing the created thing back and forth with others, as in an improvisation. Aesthetic intelligence and obsessive attention to detail help make the ball game possible. These last two dimensions of creating things that matter are what Orson Welles brought to film, Herbie Hancock to jazz, and Steven Sondheim to musical theater.

My next stories of leading artists and designers show how these last dimensions emerge and grow within creative lives today.

Aesthetic creating encompasses pioneering innovation of all kinds. We do it because we want to. It rewards our brains as our brains want to be rewarded. It empowers us with sharpened ways of thinking and feeling that not only make us better at creating, they actually make us happier doing it.

Reimagining Theater—Aesthetic Intelligence in Action

Diane Paulus is the Terrie and Bradley Bloom Artistic Director of the American Repertory Theater, creator of the Tony Award–winning Broadway revivals of *Hair, The Gershwins' Porgy and Bess*, and *Pippin*, and pioneer of a form of contemporary theater that toggles between Broadway and a planetarium, a disco club and urban streets and alleys. Her stage performance creations, in theater as in opera, have the rare ability to bring us together in large crowds, even as they invite us to peer into what pulls us apart, like prejudice and the abuse of power.

Paulus has hair falling straight around an open compelling face. Her inquiring green eyes invite conversation, as if she knows you have something important to talk about and she is just waiting for you to figure it out.

Paulus grew up in New York City, four blocks from Lincoln Center. Her mother is Japanese, and her father, American. Being near the center of the American experience and actually caring to add to it, even improve on it, has been a theme of her life. She once thought about pursuing a career in politics, perhaps running for mayor of New York City. Instead she turned to doing what she really loved: creating fresh engaging theater, though she never forgot that whatever she created would belong to society. Impossibly busy now as a mother of two young girls in Manhattan, leading a major regional theater at Harvard University in Cambridge, with ongoing Broadway productions, Paulus is warm, direct, and efficient in conversation.

For young Paulus, who danced as a girl with Balanchine and Baryshnikov (her sister played a gold harp in the center of their apartment, and her father ran fine arts programming for WCBS), the arts were everywhere except in the dirty and dangerous streets of 1970s and early 80s New York City. That absence seemed wrong, and Paulus wanted to change it. How precisely she would do that remained unclear, although she had a few ideas. Once the draw of the mayor's office wore off, she enrolled at Harvard College in the mid-1980s, where she homed in on a path in the performing arts. She started spending time with a classmate, Randy Weiner, who was a friend from New York. They had each gotten into Harvard and saw each other socially on campus. Weiner, ironic and wise, who grew up playing sports and went to Harvard thinking he was going to become a neurosurgeon, was of a completely different temperament and seemed to have unlike interests. Paulus had her path, and Weiner had his, which made it easy for Paulus to talk about her experiences and impressions. "We would meet at the end of the day," Weiner explained to me one night at an American Repertory Theater gala dinner, seated at a table with Paulus, whom he eventually married in 1995. "Diane was always fired up. She'd read some new text, seen a performance, or acted in some new production on campus, and everything excited her, made her bigger. Meanwhile, I worked away on my biochemistry. I was like, this is nuts. She's happy. I'm not."

By the time Paulus finished Harvard College, she had decided to pursue a career in theater, and Weiner had opted to follow.

Initially Paulus wanted to be an actor, imagining a rich and dynamic theatrical scene similar to the nineteenth-century Parisian scene of Sarah Bernhardt or the early American television scene of Charlie Chaplin. However, by 1988, theater audiences were already dwindling. Broadway was on course that year to set a box-office record, but attendance was down, reflecting a trend that had been going on for decades. In 1930, about 65 percent of the US population had attended a movie theater each week. By 1988, the rate had fallen to around 9 percent. It was not just that people went less frequently to the theater. In the era of modern technology and conveniences, people had fewer reasons to come together at all. That was as bad for theater as it was for New York City, in the midst of a record-setting year for murder. Paulus came back anyway. Luckily, Paul Sills, founder of the Second City, was starting a new acting school, the New Actors Workshop, with Mike Nichols. Paulus auditioned and got into the first class.

"We played endless hours of theater games," she says. This meant minding the ball in the Viola Spolin sense of improvisational theater. It meant winning and holding audience trust and creating a common experience of expressive experimentation and learning.

For two years Paulus studied with Sills and Nichols, and from there she went on to start her acting career at Off Broadway's Pearl Theater.

Almost immediately it felt wrong, and inauthentic, a sense amplified when her agent asked her to curl her hair to avoid the "Vietnamese bar girl" look. It was frustrating. Paulus loved theater and had plenty of acting talent, but she wanted freedom in what she did, and she wanted to give other actors she worked with freedom too. Actors, in her view, needed to have an authentic conversation with their audience.

Paulus called some friends. "Let's create a theater company," she announced, which led to a first gig performing Grimm's fairy tales in a community garden on the Upper West Side.

"I was interested in freeing theater from the boundaries of traditional spaces. I wanted to take theater to the streets—to bring it to a community where everyone had access to it."

Before long, Paulus sat down with Sills to talk over how she might make a career directing theater. Where should she begin? Sills had a farm in Wisconsin's Door County, which he had started to use for theater game workshops. She should go there, he proposed. So over the winter holiday season of 1991, Paulus flew out to meet with the local chamber of commerce. With Sills's encouragement, she proposed creating theater for the community in Door County.

Door County, a gorgeous peninsula in the upper reaches of Lake Michigan, was a popular summer vacation destination, and Paulus argued that a new theater group could be in partnership with the local community. By the summer of 1992, when she returned with actor friends from New York, she had raised $30,000 to start the Blue Circle Theater. The following year, Paulus directed one of the first rock-and-roll versions of *The Tempest* called *Prospero's Revenge* with her new theater group, Project 400.

The production of *Prospero's Revenge*, written by Weiner, came about in a surprising way. Had Paulus been in New York City, she would have turned to actors and musicians she knew and created musical theater geared for audiences she had grown up with. Here in rural Wisconsin she wasn't sure where to begin. One evening, while Weiner and Paulus sat at a local bar listening to a popular Door County band called Big Mouth, Paulus had an idea. During a break in the live music, she walked up to the lead guitarist, vocalist, and songwriter Jay Whitney and introduced herself. She and Weiner wanted to create a new musical. Perhaps Whitney and his group would be interested in writing and performing the music. Whitney loved the idea and said he would talk it over with his band. The next time Paulus and Weiner returned to the bar, Whitney had disappointing news. The band wanted to work with Paulus and Weiner, but the members had no time. They had young families. Their wives needed them to spend more time with their kids, not less. Paulus asked if the band members' children were old enough to perform. She thought it might work to feature the children in *Prospero's Revenge*. Rehearsals could take place during the band's early sets at the local bar, so they wouldn't need to lose time away from home. Whitney took this idea back to the group, the strategy of Diane Paulus won over the wives, and the theater project advanced.

Paulus had not only gone to a place she had never been before; she had listened and adopted an authentic voice, and as a consequence, *Prospero's Revenge* turned into her coveted cultural conversation. The rock musical became a phenomenon that summer, and its songs became local pop hits. An original thing was happening in Door County, and it caught on. Long before *Prospero's Revenge* appeared on stage, it had a Door County following, and when it finally opened, the public flocked to it.

Shakespeare's *The Tempest* reached a new audience because Paulus gave it a culturally vibrant voice. She used a form that mattered to retell a universal tale.

Aesthetic intelligence, the sixth dimension of creating what matters, is the ability to recognize and interpret aesthetic language, and with it to express new ideas that move others. We notice it in the mot juste of the politician who helps us think against the grain of political instinct—or in the political push of American electorates to the right and to the left with strategic words like "freedom," "fanaticism," or "fake."

Irony often illustrates aesthetic intelligence. In Paris, where my kids grew up, discipline, order, and hierarchy frequently framed their social and educational experiences. At school they addressed their teachers with the formal "vous" in the same deferential way with which, when adult friends came over to the apartment, my kids would come out of their bedrooms and, with an outstretched hand (or a tippy-toed kiss), say, "Bonjour, monsieur," "Bonjour, madame." My sons used irony like a salve. Irony helped them criticize and relax, express what they actually thought, and join a kind of underground of adolescent perception. When they moved to the United States, my three boys entered a very different social milieu. Here they called their teachers by first names and carried cell phones into and out of the classroom (personal electronics had been banned from their Paris school). Many months after the move, my youngest son, Thierry, said to me one evening, "My friends don't laugh at my jokes. In Paris I was funny. Here kids just stare at me." In my son's new school, irony was less comically relevant, and when it appeared, it belonged to a culture he didn't yet understand. *South Florida Wildlife Rescue Unit*, a work of contemporary art by Mark Dion, illustrates aesthetic intelligence in a similar way. Dion's yellow bus installation portrays a fictive "rescue unit" designed to head into the Everglades and save wildlife threatened by human civilization. His work is in part an ironic portrayal of policy maker attempts to curb inexorable declines in biodiversity. Created in 2006 for the art community of Miami, Dion's *Wildlife Rescue Unit* came to Boston as part of Dion's first US retrospective at the Institute of Contemporary Art in the fall of 2017. Walking his students from Columbia University through the exhibition (on a visit to which I tagged along with my son Thierry), Dion stopped in front of his yellow bus, which had an open window like a food truck (it displayed vials, cages, and old books), and said, "Contemporary art audiences used to be small, informed, and of a relatively common culture. They understood my ironic purpose. Audiences have expanded. They don't necessarily see it anymore."

We use aesthetic intelligence to adapt when circumstances change around us. In a static world, nothing is new, even if it seems to have been created. Created things in an unchanging world just prolong what existed before. In a "real" world that does change, but in relatively minor ways, created things may improve functionality, or somehow look a little better—as in the iPhone X, relative to the iPhone 8—but they don't fundamentally change our minds or our lives. When designing for such relatively unchanging worlds, creator value lies in understanding the way things work and have worked, in knowing how to make them work as well or a little better, and in marketing.

In a fast-changing world, or in a world of frontier experience, creator value lies in listening and watching, in observing new conditions and knowing how to express original observations. Creators who shape a fast-changing world in ways that endure find an aesthetic language that allows them to participate in a conversation that not only explores what might be done, but also leads to creations that are new and needed. A pioneering creator may imagine a new environmental policy, a form of architecture, or a poem. She can't know if or how this thing she has made will matter to others until she shares it. When she does share her idea, the very newness of it can make it appear to be uninteresting, useless, a distraction. But if she expresses it with aesthetic intelligence, a conversation can begin. Her created thing is noticed, consumed, remarked on. Aesthetically pleasing, it incites interest, and this interest brings a closer inspection, which eventually gives the creator insight she would not otherwise have had. After all, it is hard to ignore what is enchanting.

Our brains develop aesthetic judgment over the course of our lives. Some of this is hardwired—like sensorial isolation (as when we see a solitary human figure alone on a bridge), the grouping of forms by sensorial associations (as when we combine the tastes of Gruyère cheese and fresh grapes), or inherent properties (like symmetry and contrast). More generally, however, what catches our eye turns out to be contextual. Scientific findings of brain imaging studies across the perceptions of sight, smell, taste, and hearing show that what we determine to be beautiful is guided by what our brains associate with positive or negative valence (intrinsic attractiveness) given our particular place, time, and biology. Whether it be the sight of a painting by Georges Braque or the taste and smell of a hot fudge sundae, perceptions of beauty excite similar regions of the brain.

Diane Paulus honed her aesthetic vocabulary over five Wisconsin summers, step one of the Creator's Cycle. So successful did she become that art patrons in the Door County community eventually wanted to convert an old cheese factory to give her the security of a permanent theater. But Paulus wasn't ready. She had adapted the expressive language of the world's preeminent dramatist with a contemporary Wisconsin vocabulary, but she had yet to create an expressive language that was her own (step two).

She applied to enter Columbia University's theater program, led by Anne Bogart and Andrei Serban. Bogart was a pillar in American theater, cofounder of the Saratoga International Theater Institute with Tadashi Suzuki, creator of the Suzuki Method of Actor Training. Bogart also served as artistic director of the SITI Company and taught a famous technique of improvisational ensemble building that blended Mary Overlie's Six Viewpoints of performance with the Suzuki Method. When

Paulus arrived at Columbia in the fall of 1994, Bogart became her second mentor, after Paul Sills. Through improvisational games, theater exercises, and an immersion in contemporary theater with exposure to the Suzuki Method, Bogart helped Paulus build an aesthetic language that emphasized physical expression, to which Paulus had been drawn since childhood. Another influence at Columbia was the Romanian theater director Andrei Serban. Serban had grown up in postwar Romania, where he had caused a scandal with a daring version of *Julius Caesar* in which the audience was seated under a bridge and watched Caesar being killed in slow motion. The experience shocked Romanian audiences, who wanted to forget the trauma of their war-torn lives. It was exactly the kind of experience Paulus wanted to achieve in her theater.

After graduating from Columbia, Paulus and Weiner, now married, set to work on a new production. It was a modern staging of *A Midsummer Night's Dream*. Paulus and Weiner were driven to pull up a storyline that they felt was often lost on contemporary audiences—Titania makes love with a donkey; she is out of her mind and trapped in a kind of unnatural crowd-sourced passion. Paulus thought of Studio 54. The legendary New York disco club, hangout of megastars of the era like Michael Jackson, Andy Warhol, Elizabeth Taylor, and Mick Jagger, remained a fresh memory in New York twenty years after it had closed. She and Weiner decided to transpose Shakespeare's play from the woods of Athens to the New York discotheque. Oberon, the fairy king, became a nightclub impresario. They used famous disco music to convey the crowd-sourced passion that swayed Titania to do the unthinkable. Weiner studied the songs they'd grown up with in the late 1970s, like "Car Wash," "We Are Family," "You Sexy Thing," and "I Love the Nightlife." The lyrics, he decided, worked perfectly. Nothing needed to change. Paulus dressed her fairies in skimpy sequined disco outfits. The show came together on the Lower East Side of New York City, opening in a storefront theater called the Piano Store (step three).

The Donkey Show opened in Chelsea at the Club El Flamingo on August 18, 1999, for a limited six-week engagement. A thousand shows later, in November 2002, *Playbill* celebrated the production as a cultural phenomenon. Performances sold out for years. People didn't just have fun dancing to the disco music. The magic potion fairy, Puck, flew along on roller skates. Lysander climbed a pole and leaned over the crowd. Dimitri landed on your table and started to dance. Oberon, dressed up like an aging Andy Warhol, stalked the club, contributing to a sense of looming danger, as the fairy queen, Titania, who looked like a powerful dominatrix disco diva, ran from the arms of one fairy to the next. It was an hour of dizzying action beneath the thump of 1970s disco music. Years later, with my students, Irene and Shirley, I would experience the show with an excited audience, bound in complicity before the staging of a forbidden act. *The Donkey Show* tugged at our emotions through its rawness and originality. "The show is an exercise in breaking theatrical boundaries," wrote Lenora Inez Brown in *American Theatre* in January 2002.

The Donkey Show continued for six years, traveling

across Europe. And then, in 2009, when Paulus was made artistic director of the American Repertory Theater, the show came to Cambridge, Massachusetts. Paulus had recently directed the Tony Award–winning fortieth-anniversary revival of *Hair*, the 1968 Broadway hit rock musical that brought the hippie counterculture movement to the stage. Paulus's *Hair* used the same aesthetic language as *The Donkey Show*. "*Hair* seems, if anything, more daring than ever," wrote a reviewer in *Time* magazine.

Up in Cambridge, *The Donkey Show* continued selling out. It ran every Saturday night from 2009 through to our visit in 2017, placing it in the realm of *The Mousetrap*, a stage theater whodunit and London's longest-running show, open for sixty-five years, or the musical *The Phantom of the Opera*, running for thirty years.

The Donkey Show simplifies, shortens, and reinterprets Shakespeare's *A Midsummer Night's Dream*. It favors contemporary purpose over fidelity to the play's original content, bringing an old form to new audiences. In this sense, it is like one of the disruptive innovations Harvard Business School professor Clay Christensen writes about in his classic *The Innovator's Dilemma*. As Christensen points out, the pocket calculator disrupted desktop calculation even as, relative to desktop calculators, it diminished functionality or content. The first pocket calculators, like the Sinclair Cambridge, made it possible to perform simple calculations and store limited amounts of data. Relative to popular desktop calculators—such as those produced at the time by Canon and Texas Instruments with their hundreds of transistors—the pocket version processed calculations slowly; some operations, like those involving transcendental functions, were actually inaccurate. Pocket calculators disrupted the calculator market by addressing a contemporary need of portability and affordability while delivering an essential calculation function—however less effectively than by traditional means. Disruptive innovations, according to Christensen, seize a new need, hack what meets an existing need, and pioneer the future.

Diane Paulus brings this sort of disruption to theater. Her work has many corollaries in contemporary culture, from hip-hop to rap and contemporary art. An example is a work of contemporary art called *Rain Room*, created by the London-based artist collective Random International.

Rain Room opened at the Barbican in London in 2012 as a semicircular corridor filled with falling rain. As visitors walked through it, the rain disappeared around them. They didn't get wet. They walked along or even ran forward, staying dry through the rain. As soon as the exhibition began, visitors formed long lines waiting to get in. Soon, enough people were doing unexpected things in *Rain Room*. They danced. They sang. They ran. They proposed marriage. *Rain Room* seemed a work of contemporary art masquerading as improvisational theater with droplets.

The insatiably curious curator, Hans Ulrich Obrist, invited the three founding artists of Random International to speak at London's Serpentine Gallery. What did this artwork mean? Nothing like it had happened before. Other works of British art had caught the public imagination, such as Damien Hirst's *The Physical Impossibility of Death in the Mind of Someone Living*. But *Physical Impossibility* was an object, a decaying shark suspended in formaldehyde. *Rain Room* was, by comparison, as ephemeral as a mirage. It was pure experience. It was nature, edited. It belonged to our time.

When *Rain Room* opened at MoMA in New York in 2013, the lines grew even longer than in London. People would spend twenty-four hours waiting to walk through. Twitter and Instagram lit up with testimonies of giddiness and exhaustion. In Shanghai, the same thing happened, and, then, at Los Angeles County Museum of Art (LACMA), in the midst of a several-year drought, the buzz took on a new tone. Parents brought infants who had never seen the rain in their lives and whisked their little bodies rapidly through the falling rain in an attempt to get them wet. People wept.

Critics have attacked *Rain Room* as spectacle. It is entertainment. It is farce. Meanwhile, over a million people have filed through *Rain Room*, one by one.

The idea for *Rain Room* came about in a remarkably similar way to Paulus's disruptive idea of *The Donkey Show*. It was 2008, and the three young founders, two Germans and a Londoner, Hannes, Flo, and Stu, were building a successful industrial design firm. They had been preparing for such a career ever since their first encounter at Brunel University, outside London, near Heathrow, where they'd formed a bond and an aesthetic vocabulary shaped by a combination of engineering rigor and design precision. Later, at the Royal College of Art in London, they'd developed their expressive language and a habit of hanging out at art fairs—in Miami, in Valencia—showing their fabulously technical installations mostly for the fun of it, such as a paint roller that produced an ink drawing of a famous face, like that of John Lennon.

Hannes Koch, Random's tall, confident creative director, has a knack for thinking as he talks. Next to Koch, Florian Ortkrass is reserved, ironic, and whip smart. He speaks with wisdom and subtlety, letting Hannes air his thoughts, until, all at once, there is something new for Florian to say, which throws an utterly new light on the matter. Stuart Wood, who in the days of the *Rain Room* genesis was still among the Random founders and has since moved on to a design role at Apple, is the technical wizard, the artist who makes engineering wonderfully sensorial, a serious craftsman with less obvious patience for the messy distraction where Hannes and Florian find and shape their new ideas.

In their first years, they did not see themselves as artists. Their young firm excelled in unusual technological exhibitions that served the interests of big corporations, like the BBC. The three had decided early on to call their firm "random" as the word, which didn't exist in German, seemed to them a liberation. They had never been too sure what they were to become, and the name avoided commitment. In 2008, they had won a commission to create a giant room-size printer for the BBC, which would visualize the printing process by allowing people to see the ink flying through the air like raindrops onto the printed surface. They were working on the printer in their studio and also exhibiting a new work, called *Audience*, at Art Basel.

Audience had come about through a commission from British choreographer Wayne McGregor, who had seen their paint roller piece exhibited at the RCA and thought they might be able to create something to accompany his dancers. The Random founders made little robots with reflecting mirrors that swarmed around you when you walked by, pointing their mirrored faces at you like pigeons in a city park. *Audience* had shown in London with the McGregor dancers, and at Art Basel, it was curated by the Carpenters Workshop Gallery. Nobody expected it to sell.

The first morning of the exhibition, Hannes, Florian, and Stuart decided to go for a walk. By noon, when they returned, *Audience* had sold twice.

"You could have a career doing this," quipped the art curator, Natalie Kovacs.

Hannes wasn't sure. Paying the salaries of three artists with high costs of technical production would mean "fighting above our weight," as he puts it. They were thrilled by the sale, and Kovacs's remark was flattering. But they still assumed they would make their real money designing installations for large corporations. Once back in their studio, they continued to fiddle with the big printer that was to shoot water from the ceiling onto a surface that would instantly change color where each drop hit the ground. That had been their idea, but staring at the contraption one day, the Random founders admitted it was not turning out very well. They were working too hard to get something printed. It wasn't worth it. They started to wonder how they could salvage their work.

Flo had the first idea. "We should have people walk through this. That would be interesting. Make this about water falling on people. Not about water falling on the ground." Flo was turning to the aesthetic language of *Audience*. He had in mind a swarm of droplets that noticed you. The idea delighted Hannes and Stu, but it had nothing to do with the BBC. They needed a sponsor. Maybe the collectors who had bought *Audience* would find it of value? They did some work to reprogram their printer, then showed it to the art collectors Maxine and Stuart Frankel. A room of falling rain intrigued the American couple, and in 2011 they commissioned the first *Rain Room* for the Frankel Foundation in Bloomfield Hills, Michigan.

Random International became an art collective through a kind of discovery, just as Diane Paulus discovered she was to be a theater director. Random's founders went on to create their most enduring work almost immediately after having found their aesthetic language, as did Paulus.

They continue to explore the human-machine interface. Their machines are alive in the peculiar things they do. They incline when others walk past, mirror behavior, and twist around in surprise. They stop and watch us, and we see ourselves freshly.

Like Diane Paulus, these artists don't just see what is new and meaningful; they help us see it too.

Reimagining Design—Aesthetic Obsession in Action

Neri Oxman is an Israeli professor of architecture and design at the MIT Media Lab. Like the other creators in this book, she is passionately curious, empathetic, intuitive, frequently innocent, humble, and intelligent. As her story reveals, Oxman is also obsessive in the pursuit of her vision, a last dimension of aesthetic creating.

Neri Oxman's work, which appears in fashion shows in Paris, design shows in Milan, exhibitions in Tokyo, and museum collections in New York and San Francisco, explores the possibilities of natural function in the design and architecture of objects like clothes, shelter, and furniture. The mimicry of natural forms has, of course, an ancient tradition in architecture and design. Designers have long created buildings in the shapes of caves, trees, and hills, clothes like skins and shells, cars like horses, and boats like dolphins. They created things to last by studying what endured in nature. Now, with the help of modern science and technology, pioneering designers and architects mimic natural function as well, making buildings that regulate energy, water, and waste almost as living bodies; clothes that monitor our vital signs; cars that drive themselves; and other created things that both look and behave like nature.

Oxman is at the avant-garde of this movement. She even goes beyond a pure mimicry of natural form and function to explore how we might improve nature with new functional forms. Her passion places the implications of her work beyond our time, in a future most of us have yet to imagine. Her recent designs include a pavilion made entirely by silkworms, which she trained; an architecture for bees, to provide a stable and adaptable habitat when nature no longer can; and clothing that looks like organs turned inside out and which facilitates organ functions, like oxygenation of blood and digestion.

Oxman studied medicine at Hebrew University before switching to study architecture in London, finishing with a PhD in the Media Lab at MIT. Along the way, she spent four years in the Israeli military. With thick black curly hair and dark mischievous eyes, she grabs attention, from magazine covers to the halls of MIT. Introverted and private, she doesn't shy away from attention in the path of creating and even draws energy from it. She seems to learn by being impeccably attuned to aesthetic language; intrigued by the elusive idea of beauty, she is brave in her willingness to turn her back on it in her exploration of its contemporary boundaries.

Around the time the Random International artists started to explore the human-machine interface, Oxman began her doctoral work on the human-nature interface, intrigued by its porosity and contemporary evolution. Mashrabiya are intricate barriers that appear on second-floor facades of buildings in the Middle East. They transmit light through elaborately carved wooden lattices. Remembering the mashrabiya of her childhood, Oxman had the idea to design such a screen based on fractal patterns found in nature. She used digital fabrication to realize a screen of milled acrylic that she called *Fibonacci's Mashrabiya*. The screen poetically modulated light and filtered air. *Fibonacci's Mashrabiya* was not intended to be actually used. It was an interesting hypothesis, a research result that led to her future work, drawing on her childhood, on her many degrees, on her life in different countries, and on her fascination with myth and the furthest limits of time.

From her doctoral work, Oxman moved on to other kinds of screens, now for the body. Commissioned by the Centre Pompidou to make a series of 3-D printed plastic designs, Oxman created her *Imaginary Beings: Mythologies of the Not Yet*, a series of eighteen objects for the human body produced by the Israeli 3-D printing company Stratasys. They are kinds of screens that cling to the head, wrap around the waist, or drape over the shoulders. In the series is a helmet inspired by the Greek myth of the Minotaur, half man and half bull, designed to be shock absorbent, and intricately precious in the way of her doctoral thesis designs. Oxman used medical imaging techniques to digitally map facial features and design soft lamellas she calls spatial sutures. *Medusa*, a riff on the Greek myth of the beheaded monster, is another helmet that envisions brain-augmented electrodes for better cognitive performance. A series of *Pneuma* structures go over the chest and ribcage, a kind of lung corset that might expand and contract with breathing and help control it. *Arachne*, the weaver who gets transformed into a spider in Greek mythology, belongs to the *Pneuma* series, a stunning royal corset and lung-augmenting structure that Oxman sees as potentially increasing human performance. *Leviathan*, *Talos*, *Kafka*, and *Dopplegänger* take the series in other unexpected directions. The entire series is now in the permanent collection of the Centre Pompidou in Paris.

Does it matter? Oxman gets this question a lot.

After a recent visit of MIT Media Lab sponsors, Oxman came to give a talk to my class at Harvard. Exhausted from months of travel and several days of intense scrutiny during the annual industrial partner visit, she opted to speak from a chair next to the screen. A few weeks before, she had exhibited at the annual design show in Milan. She had shown giant works of glass printed with her 3-D glass printer. Just before, she had been in Tokyo discussing how her 3-D additive design, as she calls it, might be used to make buildings one day. She discussed these projects, while my students listened, computers closed, paying rare undivided attention. She told them they didn't need to be bound by what nature had done, or did, because nature always surprised and changed, and we could outthink it today, help it adapt to less beneficial change that human living had inflicted. As she spoke, she showed images of her silkworms at work, designing what they never would have done themselves. Oxman had distributed the silkworms over a silk fiber scaffold constructed by a robotic arm so that they would create, by filling in space between the robotically placed silk fibers, a silky pavilion. From the worms, she went on to talk of her project with bees in a nearby lab. She said she wanted to figure out how she might make bees survive their otherwise inevitable extinction.

One student, named Joaquin, asked Oxman if any of her designs had ever made it into commercial production. Wasn't 3-D printing limited to small scale and long times of production? Oxman stood up. It had been a recent topic of conversation with the MIT Media Lab industrial partners. What, finally, did it mean to Neri Oxman to have her work matter? Yes, she replied to my student, if she wanted to get her work into practice right away, she might. There was an industrial partner from the MIT consortium ready to make lamps out of her 3-D glass printer. She said they could be selling product by the end of the year. There had been an investor interested in making a high-end sporting goods company that developed her corporal designs for extreme sports. Such practical things were possible, but she wanted her designs to matter on a grander scale. They might clothe people once they went on to live in space, for instance. They might become the architectural fabric of future cities, constantly adapting to our changing needs in response to millions of living inputs, in the way her silk pavilion had changed through the agency of silkworms. They might be used to make beautiful 3-D printed edible packages for food once people cared enough about their ecological footprint. Oxman was not ready to stop creating before the sustainable future she envisioned came to pass. It didn't matter how long it took. She was living in a rented apartment and traveling nonstop. Why did she need to live like this? She actually asked the question aloud and finally left the lecture room with my students awaiting an answer. The implication was that her work mattered more than she did, or in a way it was her.

Aesthetic obsession is an instinct that creators have to support the survival of an idea. Until the idea is utterly perfect, they feel unsatisfied. Meanwhile, there is another kind of satisfaction. This comes from the persistent development of the idea and gives the creator willingness to press on.

Obsessive creators love the moment-to-moment process of discovery. There is gratification in creating that overshadows what comes next. In his book *On Writing*, Stephen King says, "I've written because it fulfilled me. Maybe it paid off the mortgage on the house and got the kids through college, but those things were on the side—I did it for the buzz. I did it for the pure joy of the thing. And if you can do it for the joy, you can do it forever." This "buzz" or stream of pleasure, what Hungarian psychologist Mihaly Csikszentmihalyi refers to as "flow," points to the meditative benefits that the Creator's Cycle brings over the sustained course of long-term creation.

The Creator's Cycle leads not just to changes in the world around us through the things we create, but to changes in our own bodies through the processes by which we create them.

Mindfully focusing attention has been shown to improve pain tolerance, reduce anxiety, lower depression, and produce other cognitive and emotive benefits. While clinical studies of thought-based therapy generally examine meditative health benefits involving concentration on one's own body or immediate environment, obsessively concentrating on process—as occurs when a creator derives a new equation, writes a new song, or designs a new video game—is itself a form of meditation. The neuroscientist Richard J. Davidson, author of *The Emotional Life of Your Brain*, and others have shown that meditative practices can alter brain anatomy in ways that parallel those instigated by the flow of sensory information to our brains. Mindfulness meditation increases prefrontal cortex activity, bolsters activity in the emotive brain, and influences many (if not all) of the emotive and cognitive states of creating things that matter. The mindfulness movement, championed by psychologist Ellen Langer, echoes the aesthetic movement of John Dewey and gets to the most profound force that motivates aesthetic obsession.

Creating obsessively changes us. Obviously, our brains differ, and experiences in our lives will have built different realms of intuition, tolerance for risk, aptitude for concentration, and so forth. Not all of us are wired to create as does a Bob Langer or a Diane Paulus—nor are any of us as performative as we ever want to be. But our brains do change, and learning to guide this change through the Creator's Cycle is among the most privileged kinds of know-how that successful creators possess.

Neuroscientists used to believe that the brain was hardwired. Electrical stimulation of particular spots on the somatosensory cortex—a strip of cortex from ear to ear that processes sensory information exchanged between the brain and the entire body—were noticed to deliver sensations of the touch of a cheek, or right foot, or eyebrow. It seemed that after childhood, our brains simply stopped developing. Eventually, scientists discovered that the somatosensory cortex actually evolves. Not using fingers for a prolonged period of time can lead to changes in the anatomy of monkey brains. The brain regions associated with feelings of finger touch can shift to serving feelings of touch in the face. Since the first primate findings of the early 1990s, many studies have shown that brain plasticity occurs in humans as well.

The visual cortex accounts for around one-third of the volume of the adult brain. In the brains of those who have gone blind, the visual cortex begins to process hearing, touch, and even words. Studies with human subjects possessing perfect eyesight, who have then been blindfolded for just five days, show similar changes in brain plasticity, pointing to a remarkable ability for rapid adaptation and survival. Many published studies now show that synaptic networks in the brain reorganize as a consequence of even a few hours of brain activity, and change durably following persistent experience, as in years of playing piano, meditation—or, it seems, the obsessive pursuit of creating.

Creators like Neri Oxman dream of the things they will ultimately make (like her 3-D glass printer, which will one day produce glass with dimensions and forms that are today unthinkable). But from the germ of the idea to the realization of the dream (from prototype to glass printer capable of generating large-scale objects with speed and quality), they may need many years. To an investor impatient to see promised benefit, these years can seem lost, but to a creator like Oxman the experience is quite the opposite. Years of obsessive pursuit change Oxman, and produce a master.

Oxman reinforces aesthetic obsession through the steps of the Creator's Cycle. Like some of history's most remarkable obsessive creators (from writers like Edgar Allan Poe to songwriters like Bob Dylan), she frequently maintains in her ideation a personal connection to her creations. She has a large group of students, and still Oxman completes her works by herself, tying them back to her history and personal experiences—as in her mashrabiya designs. As she advances, she shapes and reshapes her creations, as if connecting what is fragile and personal to what is enduring and universal—the way she associates her work to classical mythology (Minotaur and Medusa). Finally, she seems to never finish. Oxman delays gratification of dream realization, as if leveraging her many successes into bigger bets on what the universal can be (Figure 7).

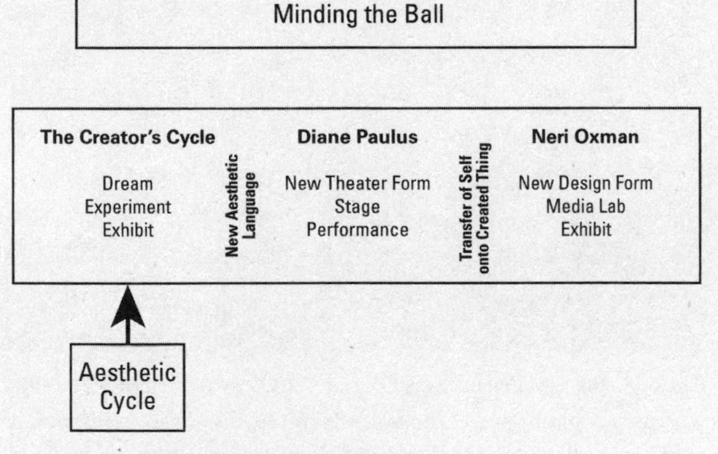

Figure 7. In the last phase of creating things that matter, the unique aesthetic vocabulary of the creator finds resonant expression in a complete aesthetic language. The public conversation that ensues seems to be a direct one with the creator herself; she is inevitably reticent to turn away. This last phase of creating can begin with the relief and clarity of a land sighting following a long, harrowing, and sometimes tedious ocean crossing—that somehow never ends.

Figure 8. *Silk Pavillion* by Neri Oxman (2013). A swarm of 6,500 silkworms was positioned at the bottom rim of the scaffold to spin flat nonwoven silk patches as they filled the gaps across the silk fibers.

Figure 9. *Fibonacci's Mashrabiya* by Neri Oxman (2009) adapts the ancient art of mashrabiya design using digital fabrication technologies.

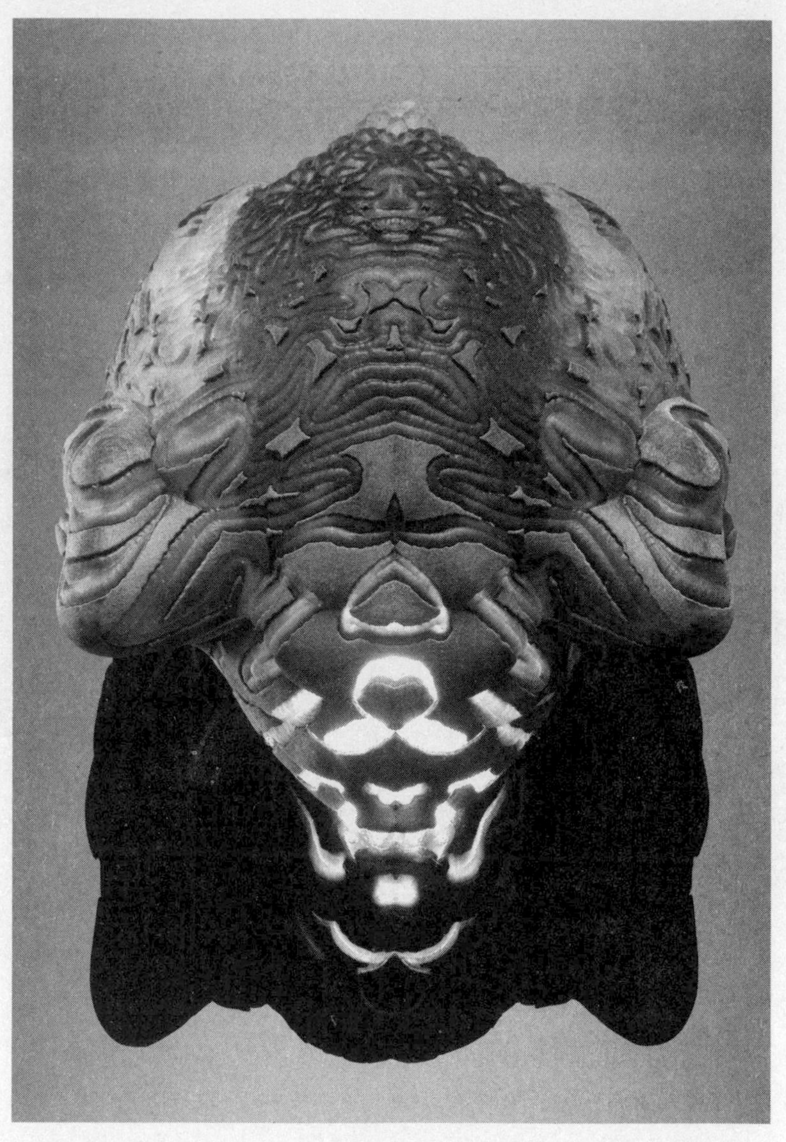

Figure 10. *Minotaur Head with Lamella* by Neri Oxman (2012) is a shock-absorbent protective helmet.

Figure 11. *Vespers Series 2* by Neri Oxman (2016) belongs to Oxman's death mask series and is now in the permanent collection of the National Gallery of Victoria in Melbourne, Australia.

Figure 12. *Gemini* by Neri Oxman (2013) was created for exhibition at Le Laboratoire (Paris and Cambridge) and is now in the permanent collection of MoMA San Francisco.

Vespers is a series of death masks that Oxman created for the London Design Museum, the Centre Pompidou, and the National Gallery of Victoria in Melbourne, Australia. The masks explore the life that begins in death, seeking to find possibilities in the regenerative potential of our last vital seconds. The Oxman death masks, made with the same processes as her *Imaginary Beings*, not only capture the facial expression of the dying, reinterpreting it in color and form; they also aim to hold the last breath of the dying and to extract from this breath living cells that will continue to live when the brain ceases to function and the heart to beat. These bacterial cells will go on to regenerate other cells through the natural reproduction process of bacteria (in the human body, bacterial cells are estimated to be ten times as numerous as human body cells themselves). With *Vespers*, Oxman continues the research of *Imaginary Beings*. Working in digital, synthetic, and biological media, with meticulous attention to detail, Oxman is shaping life itself, where it comes from, where it is going, and what (if anything) contains it.

Oxman is opening an architecture practice. She says her first commissioned work will be in Japan and of small scale. She is also in conversation with organizers of the upcoming Tokyo Olympics about how her wearables might be used to produce fluids that nourish athletes while they perform. Her future is coming, with ours.

Other renowned contemporary creators are hard at work in their culture labs exploring the future of food, medicine, and communications with restaurant goers, scientists, patients, cell phone users, gamers, and theatergoers. They often obsessively pursue their ideas, and they rarely postpone public conversation as patiently as Oxman does, waiting for all of us to catch up. What makes possible this dogged determination is Oxman's expression within the world of culture, from conferences like TED to museums like MoMA, where the pure aesthetic value of what she creates is nearly all that matters, in the sense perhaps that the drawings of Leonardo da Vinci existed for centuries as fantastical projections of what might be possible with engineering, before we fully knew the principles.

Many artists today create almost as researchers exploring an altered future. They include Jason deCaires Taylor, a British sculptor who created the world's first underwater museum, and whose site-specific sea sculptures explore a future where people revere the life of the ocean as they do the life above it; and Olafur Eliasson, a Danish-Icelandic artist based in Berlin and among the most recognized names in contemporary art, whose installations, from his *Weather Project* at the Tate Modern in 2003 to his *New York City Waterfalls* in 2008, explore a future where created things bring us back into a caring relationship with nature. Little Sun is a nonprofit created by Eliasson that manufactures small portable solar-battery powered cell phone chargers for life-sustaining electricity in the developing world away from the electrical grid. Doug Aitken is another contemporary artist whose work belongs to this movement of artist-researchers.

Aitken's aesthetic obsession explores a future where we live in mindful relationship with nature. An owl stares at the camera on a ludicrous motel bed alongside a western US highway, rumblings in the earth generate sound inside a pavilion in Brazil, and an Amtrak train full of artists creating and performing crisscrosses America, connecting us in surprising ways. The work of Aitken, whose base is in Southern California where he grew up, is a kind of random emanation from a source, as opposed to the methodical pursuit of an outcome, more research than production. In this sense, Aitken creates like Neri Oxman, but while Oxman explores the barrier between us and all that is not us, Aitken explores barriers falling down.

Following art school and a few years of living in Manhattan, Aitken settled down in Venice Beach, where he lived for ten years until, in 2010, he had a vision. He asked his parents to come to his home, which he had emptied of furniture, and sit on benches facing each other across a picnic table he'd placed in the center of the living room. Aitken then began to film the destruction of his house while his parents sat mute and expressionless facing each other. This went on for many days, in sessions of several hours at a time, as the demolition crew tore down the walls and ceilings and nearly everything else aside from the floor under his parents' feet. The result, *House*, is a transfixing work of video art, emblematic of Doug Aitken's expressive purpose.

I first interviewed Aitken at the Venice Beach home he built on the ruins of the one he tore down. He led me on a tour. "My house is an artwork," he said, "like any other work of mine." And for a while we stood under a skylight that formed the hatch leading to a rooftop terrace as Aitken discussed what light did at different moments of the day. I had come to Venice Beach to talk to Aitken about a new contemporary art prize and the possibility that he would be our first recipient. Bridgitt Evans, the cofounder of the prize, had introduced Aitken and me the day before. Over the next twenty-four months, we would meet in many other places. But our talk this day within the vanished boundaries of the *House* was the unforgettable starting point. I asked him if there might have been poignancy in what he'd done, in seeing his home falling down around his parents. No, he replied. He told me how as a boy of twelve he had sat in a car next to his dad as they drove together up into the Sierra Nevada Mountains. They listened to a cassette of Carl Nielsen's Symphony No. 5 on the car's stereo system. Close your eyes, his dad said. Listen. What do you see? Aitken saw velvet hills and crashing boulders. Then his dad changed the cassette. What about this? Aitken listened to piano keys playing faster than he'd ever heard—it was a Glenn Gould interpretation of Bach. He closed his eyes, frustrated, and confessed to his father that he was unable to see anything. His dad stared at the road as they continued along with the manic piano sounds. Aitken's dad said his son had just discovered the difference between good and great art. Good art was merely technically proficient, while great art created a sensory experience that was at last ineffable. The *House*, Doug said, had been like that. Making it or even seeing it now did not lead him to feel nostalgic or in any way self-conscious. Creating it then and seeing it now belonged to a single narrative. Where art came from and where it led was, in Aitken's view, the same place. It traced back to you, to a search for universal experience.

This third way of creating may be hard to find in our classrooms and boardrooms, but it is easy to follow if we get the chance (see Figure 13).

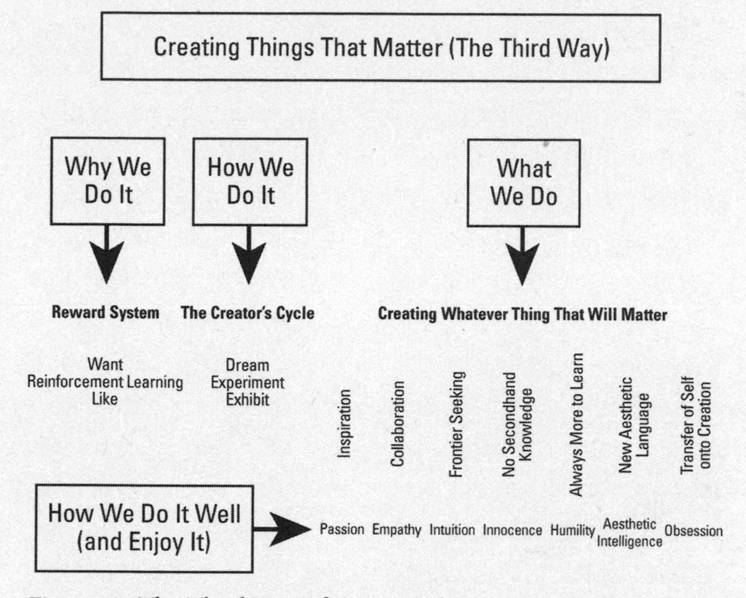

Figure 13. The Third Way of Creating, however challenging it may be to pursue in our contemporary condition, is actually the one for which we are most naturally disposed. Our brains motivate us toward it; we all cycle through the ideation-to-realization process in essentially the same way whether we are a cook or an architect. We ultimately do what our condition suggests we should do to endure. Along the way, we learn seven cognitive and emotive dimensions of creating to matter that help us do it well and enjoy the ride.

Why do we do it? The reward system of our brains.

How do we do it? We ideate, we experiment, we exhibit.

What do we do? We creatively express what matters—food, drug delivery, computers, games, theater, art.

Why is it such fun? Doing it the first time is a thrill, while doing it over and over again makes you better, so it is adapted equally to the simplicity of a child and the sophistication of an Albert Einstein.

MAKING A FUTURE WE ALL WANT

CHAPTER 6

○

Grassroots Renaissance

For the first years that I taught my class, I started each semester by sharing "seed" ideas, which my students shaped into their own. The approach kept them from wandering down the first and second paths of creating, better suited to specialized creator classes, as in an engineering design class (where you might learn how to make a new prosthetic device), a business class (where you might learn how to start a company), or a creative writing class (where you learned how to write a short story).

The goal of the class was to imagine creating something that eventually changed how people thought and lived and to pioneer the path to get there.

Over the first six years, my students pioneered a few big ideas. One was for batteries derived from wet dirt to serve areas off the electric grid. The project received World Bank funding, and after a couple years of prototypes and improvisation, it turned into a PhD project. Another foresaw soccer balls that generated and stored energy whenever you kicked them to power cell phones. This simple idea, easy to make while hard to effectively design, became the cornerstone of a company that received international fanfare. A third envisioned chocolate you could inhale into your mouth without calories. This idea led, directly and indirectly, to several companies in which many of my students became involved over the years. These were, however, the glorious exceptions. Mostly I worked with my students to help them overcome a hesitation to take the very first step of the Creator's Cycle—to dream with passionate curiosity.

The path of dreaming seemed counterintuitive to many of my students. Their classical education had prepared them for established careers, like those in finance, management consulting, and medicine, not an open-ended path of uncertainty and risk.

Then, in 2014 and 2015, my class dynamics changed. I started to receive in those years the first wave of students who had grown up in the era of portable electronic media. My new students no longer cared to follow seed ideas, and we started brainstorming their own. Before then, only a few students in each class would have had any entrepreneurial experience. Suddenly, most of my students had "maker" experience of some kind, as creators of videos, or blogs, and occasionally school clubs and startups. I no longer felt compelled to teach students what mattered off campus. In fact, they walked into class with a knowledge of what mattered that taught me as much as I taught them (in 2015 it was social connectivity, in

2016 it became reconnection from digital to material experience, and in 2017 it was electronic paths to well-being and happiness).

My students had grown up at the front edge of an expressive movement that had been underway for nearly a decade. I now view this grassroots movement as our greatest opportunity to create the world we all wish to live in.

The first tweet happened in 2006. By 2007, there were 5,000 tweets per day. By 2013, there were 500 million. A tweet is a created thing—as is of course an original sentence, a letter, or a news report. We experience something new, we notice it, drawing some observation, and we then express what we have observed. The scope of a tweet is far narrower than that of a novel, but we do create when we tweet. The vast majority of tweets are not composed with commercial intention, nor are they meant to be enduring works of culture. People tweet mostly out of a sheer desire to express themselves. Twitter expression may be angry, superficial, grammatically incorrect, and deeply hurtful. It may also be hopeful, profound, and poetic. It expresses in a limited way what one person happens to be thinking and feeling, for whatever it's worth.

What has happened in the sudden and spontaneous conversion of so many world citizens into active tweeters, in the willing participation of hundreds of millions in a kind of graffiti expression of reality, has happened to us all, or pretty close to us all. Facebook has around 2 billion active users today, WhatsApp has over 1.2 billion, YouTube 1 billion, Instagram 600 million, Twitter over

300 million. The leading blogging site today is Blogger, with over 43 million unique visitors per month. According to *Entrepreneur* magazine, millennials spent an average of 5.4 hours a day on user-generated content in 2014. Not even Emily Dickinson wrote that much each day.

Beyond the tweets, Instagram posts, and other forms of online user-generated content that infiltrates culture and commerce, there is also a parallel and growing movement of making material things often simply for the joy of making. These things, as with tweets and blog posts, may be aesthetically refined, or rough prototypes, carefully wrought, or thoughtlessly made with little sense of craftsmanship. They express who we are, however imperfectly, at this particular instant in time.

If there is a common experience of living today on the planet, it might best be characterized not by political, religious, or social viewpoint, but by the simple act of creating for expressive purpose.

Does it matter?

Angry tweets do not make the world better, nor do they matter as created things that bring enduring value to the lives of others. When the artist John Gerrard makes fluttering flags out of spilled oil films on ponds of water, he is expressing an angry thought, albeit in a beautiful way, with aesthetic intelligence—and most tweets have nothing even close to Gerrard's beauty of expression. But the inclination to sensibly experience life, to share our experience with others, and to enter into an expressive conversation—does matter. If we are to make a hopeful future for all, every voice counts.

For the first time ever, a significant swath of the general public is expressing itself in ways that endure, not just online, but in all the generated forms of what I call the Grassroots Creator Movement (GCM).

Grassroots Creating

Over the last decade, people have started to make things without being told they should and on a scale never seen before. These things include tweets, blogs, objects we carve out of wood, and other things like 3-D printed objects, startups, improv performances, quilts, tiny houses, robots, magazines, books, vertical gardens, novel food forms, BioBricks, clothing, sustainable designs of all kinds. It is as if, like early human cave dwellers, we were all putting our marks on the world for the first time.

The maker movement, the hands-on core of the GCM, began around the time of the first tweet. *Make* magazine launched a year before, in January 2005. In 2006, an exhibition fair dedicated to artisanal arts, crafts, and technology opened in San Mateo, California—the first Maker Faire. Maker Faire attracted 25,000 visitors in its first year, and by 2013, visitors had grown to over 125,000. Today, Maker Faire is in cities around the world, with an audience estimated at 2.3 million globally. With the first Maker Faire came the first open-access maker space, TechShop, a permanent venue inside of which culture labs can come and go. TechShop, which started in San Francisco, is now the largest maker space in the world, with sites around the country, including in Detroit, Los Angeles, Austin, and Arlington. Each site offers public access to the latest micro-manufacturing facilities for the cost of a gym membership. Online maker spaces, including games for making food (like *Food Street* or *Papa's Donuteria*), fictional characters (like *Fallout* or *The Sims*), or rooms (like Martin Wattenberg's *Apartment*), have proliferated, as have online micro-manufacturing and retail forums like Etsy, Indiegogo, Tindie, Kickstarter, and Dragon Innovation. By the time the first White House Maker Faire took place in June 2014—an exhibition spread over the White House complex with homemade inventions ranging from a giant robotic giraffe to a machine that made music by tapping bananas—over 150 colleges and universities around the country had created maker spaces. The commitment of companies like Autodesk, Intel, and Disney provided maker tools and settings for millions of students and for the curriculum work of non-profits such as Maker Ed. The maker/do-it-yourself (DIY) movement—a particularly visible core of the more diffuse GCM that includes every tweeter and YouTuber—has grown to around 135 million in America alone, according to the Atmel Corporation, fabricator of the microcontroller in the popular Arduino units (ready-made circuit boards) commonly used in maker micro-fabrication.

Grassroots creation drives startup culture. From restaurants to technology companies to financial service providers, startups emerge from the brains of entrepreneurs who create new things mostly because they want to. Since 1977, virtually all new job creation in the United States has come from commercial startups, according to the Kauffman Foundation. Special maker spaces for startups have consequently appeared, like the Cambridge Innovation Center, which is opening in cities around the world and provides offices, meeting rooms, and social network opportunities for early-stage startups and loops in big industry, much like TechShop, if more oriented toward the practical and commercial end of the aesthetic creation spectrum. Oculus, the crowdfunded virtual reality company purchased by Facebook for $2 billion, is an example of a commercial GCM success. On the cultural end of the aesthetic creation spectrum are enduring works derived from personal experience, cultural dialogue, and creative process. These works express in contemporary aesthetic terms new ways of thinking and living, from successful contemporary novels, like those of Zadie Smith and Ha Jin, to the music of popular hip-hop artists Kendrick Lamar and Dr. Dre. Grassroots videos have meanwhile helped radically change personal television viewing habits (the first YouTube video ran in April 2005), moving television from an exchange between corporations and individuals to more of a global peer-to-peer conversation. Reflecting this change in viewing habits is the rising influence of on-demand video content providers like Netflix and Hulu.

You might picture the GCM as a giant funnel of ideas (Figure 14). Millions of ideas enter at the top. Most are little more than ephemeral, spontaneous expressions of personal experience. Some advance in the minds of creators (they descend into the funnel) by a process of iteration and experimentation in "transient" culture labs that promote learning. People come to these more for the value of the creative process than for what this process delivers; they enter them to learn maker techniques, team collaboration, and open-ended problem-solving

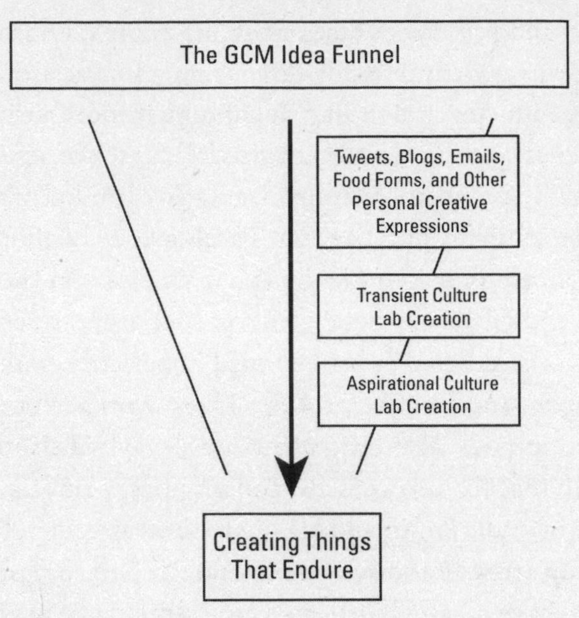

Figure 14. Over the long course of creating we are inevitably mostly mindful of our particular frontier—our place and time. Occasionally we may glimpse a universal process. The future gets created by an unimaginably complex human collaboration—the human brain is often called the most complex functioning system in the universe, while a far more complex system is the collection of brains that make up the human world as we know it. This supra-brain collaboration has grown exceedingly richer since the start of the Grassroots Creator Movement. Hundreds of millions of enduring expressions precipitate creative ideas each day, which develop in transient teams ranging from families to groups of friends and collaborators. From these ideas emerge creations that get shaped into durable created things by dedicated individuals who lead the kind of aesthetic lives described in chapters 3 to 5 of this book.

skills—along with the aesthetic dimensions of creating. They then head back to their "normal" lives, refreshed, as Ken Ledeen was when he returned from his transient culture lab in Kathmandu, a story I tell next. This role of the GCM transient culture lab as starter or refresher class in creativity has become so compelling as a path to innovation learning that there are schools entirely formed on its basis. An example is NuVu, started in 2010 in Cambridge, Massachusetts, by a group of MIT students and faculty. Without grades, without subjects, without courses, and without a schedule, experimental schools like NuVu illustrate how GCM transient culture labs frame a new way of learning suited to the rapid-changing world we live in today. Ultimately, a few ideas (appearing at the base of the funnel) enter "aspirational" culture labs that provide creators sustained resources and encouragement, before exiting the funnel as durable created things with commercial, cultural, and social value.

Why We Make Transient Culture Labs

My dad had a mentor named George. George was a half brother to my dad's mother, conceived by my great-grandmother one night on an outing to a traveling circus. Born in scandal in Columbus, Ohio, the "buggy capital of America," George was shuttled to the home of Aunt Mary, who dutifully raised him until the age of five, when he started kindergarten. One day that year, he came back from school to find his clothes on the front porch inside a bag at the end of a stick. Aunt Mary had finished her social service. George was old enough to walk to school, and Aunt Mary had decided he could just continue walking. With nowhere else to go, George ran off to rejoin the traveling circus where his life had begun.

The years passed, my grandmother stayed in touch

with her half brother, and by the time my dad turned five, his uncle George—by then a successful salesman in Chicago, handsome and streetwise—was a frequent guest in my dad's house. Among other things, he taught my dad how to create wonder, thrills, and admiration as a magician, a skill he'd learned from the circus. My dad soaked all this up. His home turned into a culture lab, an incubator of the Creator's Cycle, which lasted as long as Uncle George was in town. Dad dreamed up new tricks, experimentally perfected them, and eventually exhibited his creative output to the adoring public of his mother and grandmother. My dad excelled as a young magician and eventually turned his learning to practical use as a terrific teacher.

Long after George had died, I became curious about magic and the mysterious figure of my great-uncle. My dad had held on to some of the old stage tricks, including the Chinese chafing dish from which you could produce a live rabbit, and the magic wand that wilted when you handed it to a member of your audience. The important first step, my dad explained to me (as he introduced me to what would become my own transient culture lab), was to practice in front of a mirror. You needed to learn to see yourself as everyone else did. When you could almost fool yourself, you were ready to perform for others. During my middle school years, I spent hours each week honing my skills. My parents inevitably became a first audience, and occasionally my two sisters. Later I showed my tricks to my friends, and eventually I did shows for neighborhood kids.

Every time I did a magic trick before an audience, I

did it slightly differently. I learned to improvise by watching my audience, and I gave them what they wanted, which was usually to be fooled in a way that verged on the miraculous. This is why I never repeated magic tricks before the same people. Repetition would defy the idea that we had just created together something precious and rare, a fleeting belief in unbounded possibility.

I experimented in this way and got better as a magician, like Uncle George had while in the traveling circus, and as my dad had in his childhood home during his uncle's weeklong visits.

Transient culture labs like these are natural settings for the creating of new expressive things. By supporting the reward cycle of the creative process, they have produced global creative successes, like the Beatles and Ford Motor Company, and more subtle successes, like three generations of amateur Edwards magicians. With the growth of the Grassroots Creator Movement, these labs are now proliferating to form a relatively organized aesthetic creator learning platform.

Transient culture labs of the GCM might involve a team building a home in Guatemala with Habitat for Humanity, or food entrepreneurs getting together for a weekend Eat Retreat. Participants get involved for educational, professional, and personal development, and they often have a shared passion for humanitarian or cultural engagement. The time frame can range from days to weeks, making participation adaptable to almost any educational or professional career. People follow creative paths that mimic those of lifelong creators—pursuing projects with passion, working together empathetically, intuitively expressing themselves, and perfecting whatever aesthetic works. What they create matters to them and at least has the chance to matter to others outside the lab.

Wood Day

Soon after moving my Laboratoire from Paris to Cambridge in late 2014, I met Ken Ledeen, who gave me my first real exposure to the Grassroots Creator Movement.

Ledeen, a successful technology CEO, had been gradually retiring for years. A husband, father, and grandfather, he already led a full life. He had a wide network of friends going back to his college years, and frequently invited them over to have dinner with him and his wife, a terrific cook. He also showed up occasionally at his company, which he had founded many years before, even though by then it ran just fine without him. In his spare time, Ledeen could often be found in his woodshop at his home outside of Boston. There was something transcendent about working in that wood shop, and Ledeen had a passion to figure out what it was.

Ken Ledeen enrolled in Harvard University at seventeen years of age in 1963. Nobody in his family had attended college. He studied literature while he learned to write computer code to pay his way through school. When he graduated, Ledeen had a chance to start a company. He had never imagined a career in business, let alone leading a technology company. He remembers his investors handing him a piece of chalk and asking him to draw two columns on a chalk board. One column belonged to debits, the other to credits. Ledeen quickly caught on to accounting, though it took a while longer to master leadership. Ledeen didn't have an education in computer science or engineering, a liability for starting up a tech business. Then again, he did know the works of Shakespeare, T. S. Eliot, and Robert Lowell. He knew how to write compelling narratives and understood that nothing is ever quite right in the first draft—and these things actually served him more than he could have imagined. Running his software development company Nevo Technologies was essentially a matter of leading a creative life with real stakes. "Success was about vision," he says, "about knowing how to dream up a story, tell it convincingly, and figure out how to make it come true. It was like writing a great book."

Ledeen thrived as CEO. As his children grew up and started to leave home, he set up a woodshop in his garage and began to make furniture. He had taken some woodworking classes, but mostly Ledeen created on his own, "for the therapy of it," as he puts it. Occasionally he went to seminars and workshops to learn techniques and draw inspiration. On one of those occasions, Ledeen met a Maine-based wood artist named Jacques Vesery.

"I was blown away following a presentation by Jacques. I'd never seen anything quite like it. So I went up to him afterwards and asked him if he'd teach me how to do what he had just shown us. He was gracious, etc. etc. Ultimately, he said he really didn't have the time. I was, to say the least, persistent. Out of a bit of desperation, I offered our Paris apartment as a trade; one day in Paris for each day of instruction. Finally, I had to threaten to tell his wife about the Paris vacation offer."

The two became close friends.

Jacques Vesery—modest, talented, caring—is the kind of leader who makes the GCM the explosive force for change that it is. He lives in a beautiful wooden house in the middle of a Maine forest. From the front door, you can access his woodshop in about as many steps as it takes to make it into the kitchen. An impressive collection of tools is meticulously displayed along the woodshop walls, within easy reaching distance. Above the tools are carved works of art, exquisite, tiny sculptures that show such fine attention to detail that close-up photographs can make them appear ten feet tall. Near his workbench is a repurposed dentist chair where he does intricate carving.

Vesery's work is widely admired, and he is popular on the artist talk circuit, which is why his name came up when organizers of "World Wood Day" decided to host a second international maker retreat.

World Wood Day commemorates the ancient art of wood-making by an annual celebration on the vernal equinox, March 20 or 21, in different places around the world. In 2013 Vesery traveled to China, where he joined a group of international wood artists to create a work of wood art. The experience proved delightful, and he was asked to lead the celebration the next year, in a remote town in Turkey called Kopru.

Twenty-three wood artists joined him. They had no common language and hardly even a plan. None of this deterred Vesery. He split the creator experience into three phases, intuitively following the steps of the Creator's Cycle. He wanted the first phase to take place entirely on day one. The goal was to get everyone ideating together to figure out what exactly they would make and how they would make it. Vesery encouraged brainstorming, making sure every idea was shared and discussed by the group. Artists made models to illustrate their ideas, advancing personal visions, but by the afternoon they'd arrived at an impasse. There were two camps of thought. The goal was to make a bridge. One saw it as a classical form, the other as more contemporary. They couldn't decide, and late in the first evening, Vesery called it a day. They'd missed their first deadline. A night's sleep helped; the group came back together the next day wanting to get on with the project they'd traveled so far to complete, and by midmorning everyone had agreed on the contemporary form. Vesery marked this moment with the gift of a bunch of sky-blue T-shirts he'd designed since arriving in Turkey. Laughter filled the room. The artists slipped on their shirts, which instantly changed them from a collection of individual artists with different points of view to a single creative unit. From this moment the team pulled together, becoming more than the sum of each of their talents, excited to do what they'd all decided they should do. (In the World Wood Day video that documents the two-week collaboration, this second phase of bridge building comes across as the honeymoon phase. Artists—from Europe, North and South America, Africa and Asia—describe their experience using words like "family," "warmth," "unique," and "miracle.") The wooden bridge took shape, and then, two days before installation, the artists decided they'd chosen the wrong site. Vesery led everyone through the last-minute crisis of repositioning the bridge in Kopru, and with snow falling, icy fingers, and big smiles, the artist collaborative finished its work.

"Leadership is not about telling people what to do," Vesery said to me a few years later after we had hosted a similar experience with my own students, but "helping people reach a common decision."

Ledeen came across the documentary on the World Wood Day website soon after I met him, and he decided he had to participate the next year. He was hardly an internationally recognized wood artist, but he felt he might persuade Vesery to let him join. He did, and the year Ledeen sat through my course for the first time, in 2016, he flew to Kathmandu to participate.

Arriving in Tibet, Ledeen drove straight to the village of Bhaktapur with an advance group that included Vesery and four other artists from Ireland, Iceland, Jordan, and Australia. The team researched the site and found resources, tools, and a place to work. Conditions were tough. The 2015 earthquake had been devastating, but the subsequent blockade by India had proven far worse. The water quality was poor (the advance team settled into the habit of drinking "quite a lot of terrible beer"), the food often contaminated ("we did fine, but no fruit, no salads, basically curries, all served really hot"), and the air quality terrible. Kathmandu is among the most polluted cities on earth. By the third day, everyone was wearing masks. Electrical power sputtered into workshops and hotel rooms for just twelve hours a day. It was intermittent and hard to predict.

Ledeen had come to an actual frontier. Not only did he and the others face the challenge of creating something new and valuable; they had the added worry of daily subsistence. "Our first big challenge was to find a place to work," explained Ledeen in the first conversation we had after his return. "We ended up in a school yard just off of Durbar Square in Bhaktapur. The school leaders agreed to let us use their yard, which was really a rubble heap, in return for us cleaning up the space and exposing the kids to our art making."

The entire team arrived and got settled into a set of small hotels and guesthouses around Durbar Square. As he had done in Turkey the year before, Vesery explained that everyone was to decide what they were going to build, focus on a single idea and design, and get to work on it. "The notion is that you will not point to some aspect of the final work and say, 'That's mine. That's what I did.' If you are a carver, Jacques said, then someone else will draw something; you'll start to carve it, and then you will walk away and someone else will finish, possibly changing it along the way. We allocated three days to the design process."

Kids had few options for recreation at the school—the only playground equipment was a small metal slide. The team quickly decided that their work would serve as a playground installation. Kids would be able to climb onto it and play. Vesery divided the group into four teams, handed out paper, and the process began. "By the end of the second day, everyone's competing ideas had been absorbed, digested, reconfigured, and a single vision had emerged. There is no question that this is the most interesting part of the process. From the outset, we stressed that all ideas were on the table. Not that there weren't any bad ideas. There are plenty of them. Rather, the notion is that the best ideas often emerge as the third or fourth iteration of what began as a pretty dreadful idea."

They decided to create a sculpture out of a series of triangle frames that kids could climb onto and crawl through, using carved bricks with engravings of student palm prints, as well as a wooden frame with a woven net, bamboo bells, and steps for little kids. "We started the process with an hour-long ceremony to bless the tools, conducted by the five Nepali carvers," Ledeen described. "They said that they want to sacrifice a goat and sprinkle the blood on the tools in order to protect us all, but out of deference to western mores, they skipped that. The first day, I walked into a cross beam and split my scalp open. One of the Nepali guys noted that I could have avoided that had we just done the goat sacrifice."

At last, they completed the two playground sculptures, celebrated in a ceremony in Kathmandu, and flew back home.

Listening to Ledeen, in the summer of 2016, I became intrigued by the power of his collaborative creator experience for learning the aesthetic dimensions of creating without the long-term commitment generally associated with the birth of things that matter. In just a couple weeks, he had distilled the essence of the enthralling experience that guides some to fulfill creative dreams that take a lifetime.

How Transient Culture Labs Work

Creators in transient culture labs also follow the Creator's Cycle, but they do so with a clear short-term purpose.

In the *Ideation* phase, creators embrace common objectives and rules of engagement. In the case of the 2016 Kathmandu experience, the objective was to make an interactive wood sculpture for the Nepal Academy, the nature of which would be determined by the group in the first three days. Two leaders, Jacques and Cillian, managed the ideation process, helping everyone to stick to the rules and facilitating the desired outcome within the prescribed time constraint.

Time constraints may seem antithetical to aesthetic creating. We can see the writing of a book or the designing of a home as our life's work—the created thing somehow expresses who we are or imagine ourselves to be. We sense that when the book or the home is completely finished, we will somehow be finished, and the dream will vanish in the instant that the reality appears. So long as the beauty of the reality does not stand in, durably and reassuringly, for the power of the dream, we do whatever we can to press on with the dream. Even then, we have constraints. Sometimes these get imposed by others, while experienced creators learn to impose them alone. A creator may want to complete a window design of a dream house, in order to solicit a cost estimate of a glass artist, which might allow her the chance to decide about the window and mock it up for springtime when the sun shines at a favorable angle. Her time constraint frames the conditions (the start and end lines) of a race through the Creator's Cycle, which will reward her and give her even more energy to continue on. Transient culture labs lay this same tension over a process that will go no further in time than this, and by highlighting the steps of the Creator's Cycle they build confidence that normally comes from years of experienced creation.

Ideating in these conditions drives creator passion and empathy in ways common to the examples of chapter 3. Possessing little or no experience, creators are given the opportunity to dream, to experiment with others, and to express a shared dream. Together, they share the stakes of the dream, bear the chance that it will fail, and have the opportunity to experience a good outcome. Passionate curiosity and empathy increase as a consequence.

Wood Day was an emotionally powerful experience because it used these techniques successfully, not only through the thoughtful leadership of Vesery. It also placed everyone under tight constraints and agreed-on rules of engagement. The challenging circumstances of Bhaktapur, where even the basic needs of eating, sleeping, and breathing were difficult to meet, further heightened the group's interdependence and added to the team's learning experience. Other transient culture labs do the same thing.

Rural Studio is a longstanding program at Auburn University, started in 1993, which teaches design and architecture students at the university by inviting them to conceive, design, and implement low-cost projects that meet essential needs in western Alabama. Projects over the last twenty-five years have ranged from sheds to houses and have also included a long-term project goal to make a scalable $20,000 home that is not only safe and functional, but aesthetically appealing too.

Rural Studio teams typically include three or four

students. They work together over the course of a semester and follow the three classic creator phases. Like Wood Day, Rural Studio places creators under time and collaborative rule constraints, includes challenging circumstances, and encourages team collaboration and idea synthesis.

Many other collaborative culture labs within the GCM work along the lines of Rural Studio. They can range from educational projects, such as Design for 90 at Ohio State University, to volunteer initiatives, such as the Community Design Collaborative in Philadelphia or Habitat for Humanity International, and to community-embracing design firms, such as Shophouse & Co in Singapore. Each team has a leader who guides ideation under collaborative rules and tight time constraints, often in new and challenging circumstances. Volunteers, students, and professionals work together toward an educational or humanitarian outcome. They finish the experience, wanting to repeat it, as did Ledeen. Not only is the process rewarding; success often ignites passion to do it again and builds rare bonds. The ideation phase of transient culture labs is so powerful that some labs make it their primary focus, like Science Foo Camp, an invite-only gathering of 250 scientists, technologists, and inventors each year, organized by O'Reilly Media (FOO stands for Friends of O'Reilly). The Sci Foo Camp takes place on the Google Campus in Mountain View, California. The goal is not to create but to share ideas, build passion, and forge new collaborations.

In the *Experimenting* phase, creators make things in a workshop, a kitchen, or an online collaborative site.

Leadership keeps the team collaborating around the agreed-on objective and time line.

The experimental making phase rewards the cultivation and practice of intuition, innocence, and humility, as illustrated in chapter 4. Creators consciously build their own futures by basing each new day's agenda on the previous day's results. Creators realize they can fail, and they sense the cost of failure. This promotes clear thinking and collaborative dialogue. When the time is right, creators share their discoveries and change what they are creating as a consequence of how others perceive their work (as Ledeen pointed out, "Not that there weren't any bad ideas, there were plenty of them!").

More transient culture labs are popping up and following similar protocols. Eat Retreat is a four-day gathering of food lovers, including chefs, food writers, entrepreneurs, and farmers who hone their ideas around campfires and make delicious things to eat and drink. Creating together, participants in Eat Retreat learn new techniques that help them cook better. But cooking new things over and over again as a team—pushing themselves to create and test in a short time frame—also produces a powerful experience that rewards innocence, intuition, and humility. As food blogger and two-time Eat Retreater Rachel Adams says, "It changed how I cook, how I approach food, and the way that I eat in such a massive way that it is not even quantifiable."

The *Exhibition* phase rewards aesthetic intelligence and an obsessive commitment to perfection. The 2016 Wood Day gave creators a chance to exhibit what they'd made in the Nepal Academy for kids of Kathmandu.

Wood Day creators learned a local aesthetic vocabulary informed by their own cultures and the Asian setting in which they worked. The process gave each creator a personal connection to what was being made, and by expressing their work, they began a public conversation.

Exhibition can take a variety of forms. Collaborative fiction writing, for example, engages three or more writers in creating a common story. A creative practice that has grown with the rise of the Internet and multiuser role-playing games, collaborative fiction actually traces back to the last Renaissance, where collaboration in the writing of plays and stories seems to have been more the rule than the exception. Renaissance English theater was famous for collaborative fiction writing. *Sir Thomas More* was a collaboration between William Shakespeare and Anthony Munday, as well as other authors, as were other works of Shakespeare's, such as *All's Well That Ends Well* (recent research suggests a collaboration with Thomas Middleton, who may have contributed to *Macbeth* as well). Collaborative fiction can be used for learning and community engagement, as in the 826 National, which came out of writer Dave Eggers's 2008 TED Prize—or to create an engaging ever-evolving novel. Rarely today does collaborative fiction aim to create an actual commercial work, which may largely owe to the unpredictability of the collaboration process and the unforgiving nature of current commercial markets. Forums or culture labs for collaborative fiction thrive in online sites for universe creation like *Epic Legends of the Hierarchs*, multiuser role-playing games like *Dungeons & Dragons*, and online collaborative editing platforms like Google Docs.

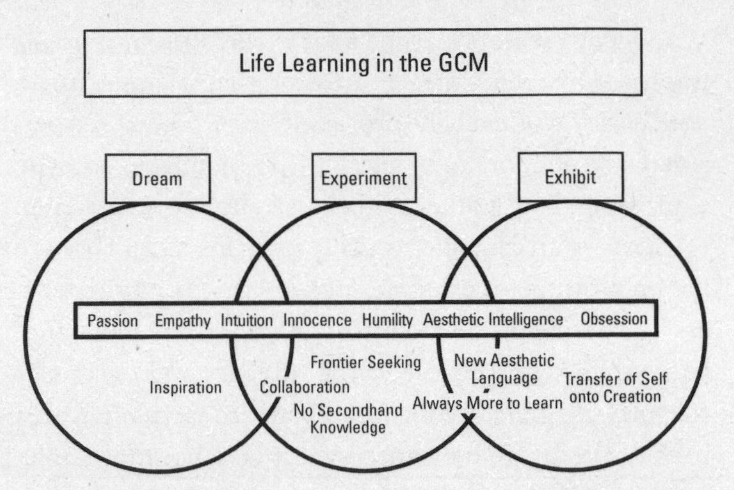

Figure 15. The GCM transient culture lab is an extraordinary model of voluntary human productivity. That's why it is starting to reshape the workplace.

Collaborative creation, however it happens, is critical in remaking the world around us. From the Pyramids of Giza to Google's search engine, human collaboration in the making and realizing of created works has always been key to achieving scale, scope, and durability. We produce films, clothing, food, automobiles, books, and countless other commercial and cultural goods by organizations that collaborate in ways that do not resemble what I have described here. In the Grassroots Creator Movement (GCM), people come together not to get a paycheck or meet a corporate deadline, but to create beautiful things simply because they want to, because it makes them happy. They produce willingly and often without any expectation of remuneration. They go to remote places, serve humanitarian causes, and develop creative skills. There are other beneficial consequences too (Figure 15).

While collaborating with others who help, teach, and inspire them, they express themselves in personal ways that build friendships and meaningful communities and learn how to push ahead in ambiguous circumstances of the kind increasingly common to work environments today.

The Grassroots Workplace

According to the Global Wellness Institute, one of many nonprofits created since the first national wellness institute formed in the United States in the late 1970s, about 38 percent of the world's 3.2 billion workers feel stressed at work, 52 percent of adult workers are overweight or obese, and 76 percent of workers are psychologically struggling. Absenteeism, estimated to cost the US economy around $150 billion, is nowhere near as concerning as presenteeism (showing up at work when personal circumstances prevent productive engagement), estimated today to cost US productivity over $1.5 trillion annually.

The crisis of workers who show up without passion, listless and unable to perform, contrasts with the culture of the GCM, where creators will even pay for the opportunity to do productive work. Consequently, some of the top corporations in the world now look to the GCM for inspiration. In the spring of 2017, IBM became the first major corporation to move its employees into an entire building in New York City run by a collaborative workplace company—WeWork. Real estate companies like WeWork, or the New England startup Workbar, have drawn lessons from the methodologies of transient culture labs of the GCM in their redesign of the workplace. They try to augment whatever facilitates ideation, experimentation, and exhibition, and to lower barriers to tasks that get in the way of creative process. Collaborative workplaces make it easy for anyone to get a functional office. Other local work spaces exist today in settings of high GCM concentration, including San Francisco, Hong Kong, New York, London, Amsterdam, and Boston, forming impressive networks of grassroots creators with diverse fields and backgrounds, from single writers to teams of entrepreneurs. While their businesses may have nothing in common, creators and entrepreneurs benefit inside these collaborative work spaces from the presence of others who value the Creator's Cycle. Collaborative spaces aim to increase impromptu encounters by transparent separations, common spaces and amenities, and the various social and networking activities such as weekly happy hours, group physical and mental activities like yoga, as well as informational activities like talks that foster ideation, experimentation, and exhibition.

The third way of creating is spreading in popularity, showing up in transient culture labs from my Harvard classroom to the IBM workplace. But these surprising and exciting developments do not alone create things that endure to change how we think and live. They require "activation" of the creator process (more on this in the next chapter). Activation transforms transient culture labs into sustainable settings for long-view creating similar to those that are home to many of our leading creators today.

CHAPTER 7

The Fire of Renaissance

THINGS THAT HAVE NOT CHANGED FOR A LONG TIME, like a dry field covered by lots of dead plant matter, exist in a fragile equilibrium with nature. Each new leaf that falls, every new day without rainfall, makes it more likely that at some point a fire will sweep over the field, turning leaves and other plant matter into accessible nutrition, like potassium, calcium, and phosphorous, and freeing up roots in the soil to feed the microbial well-being of whatever comes next. Sparks provide what scientists call "activation energy."

Renaissance activation works like this too, as it did during what we call the Italian Renaissance.

The 1453 fall of Istanbul—Constantinople at the time, once the seat of the Roman Empire, while in decline for over a hundred years—threatened Europe, not only with a possible military invasion by the Ottoman Empire. It meant European products lost markets, while the prices for products from the East, like spices, skyrocketed. To survive and maintain their former lifestyle, Italians (and other Europeans) needed to pay more attention to the world, discover maritime trade routes to the East, and build new industries. Fresh ideas for change emerged in the sciences and engineering, such as the engineering and design innovations of Brunelleschi and the scientific discoveries of Galileo and Copernicus. Innovations also famously emerged in the arts, where the works of Michelangelo and da Vinci changed perceptions of life and death. These creations happened with critical support—or activation—a process fundamental to creating what matters and illustrated by the story of Filippo Brunelleschi.

Brunelleschi is often called the first engineer of the Renaissance. Self-taught, he pioneered the field of engineering while also becoming an accomplished mathematician, artist, and architect. Brunelleschi invented the principles of linear perspective, which became the basis for pictorial representation of space through the nineteenth century. His discovery, made around 1415, helped Galileo, in 1609, figure out what he was observing when he pointed his new optical telescope at the moon, prefacing the era of modern science. Brunelleschi eventually realized his most famous work, the dome of Florence's Santa Maria del Fiore (the Duomo), through a series of design experiments. These involved architectural commissions by Florentine patrons including Bartolomeo Barbadori, Andrea Pazzi, and the ultimate activator of the Italian Renaissance, Cosimo de' Medici. Medici, a banker who managed the money of the Pope and the entire

Roman Catholic Church, funded completion of the Duomo when Brunelleschi was bankrupt.

It is easy to conclude that wealthy patrons, such as the banking family of the Medicis, alone activated the Italian Renaissance, but this misses an essential reality of how activation actually occurs. Brunelleschi's genius was activated because the following three criteria of activation were met. The first was his lifelong commitment to aesthetic creation. Second, a series of patrons (let's call them activators) gave support again and again during the creative process. And the third was the city of Florence itself, which brimmed with a culture of possibility and encouraged brilliant creators from the artist Donatello (a close friend of Brunelleschi) to Michelangelo and da Vinci. Florence was a culture lab—not in the "transient" sense of the last chapter, but in the enduring or "aspirational" sense of the culture labs I described in chapters 3, 4, and 5.

Renaissance activation finally works like fire. It requires committed aesthetic creators, who are like the *dry kindling* of creating things that matter. Activators behave like *sparks*, and aspirational culture labs provide the *oxygen* by which sparks on dry wood can grow into fires.

Grassroots Activation

The Florentine patrons might have spent their money on less risk-taking creators than Brunelleschi and furthered the already handsome architecture of the fourteenth-century city. This would have avoided the chance that

Brunelleschi's new self-taught methods simply might not pan out. They bet, knowingly or not, on the ability of Brunelleschi to realize the most ambitious engineering and architectural dream of the Renaissance, the Duomo. What they got in return was the opportunity to experience things of exceptional beauty—and a chance to enter history at a time when leading change meant not being swept aside by it. The Florentine cathedral had actually been designed in the late thirteenth century on the ruins of an ancient cathedral by Arnolfo di Cambio and constructed over the fourteenth century with oversight from Giotto to Giovanni d'Ambrogio. Construction was nearly finished by the early fifteenth century. All that was lacking was the cathedral's giant dome, which was to be larger even than the ancient Roman Pantheon. Nobody knew quite how to make it. The intuitive Brunelleschi won a commission to build it for the city by cutting out the bottom of an egg and standing it up on its cut-off base to demonstrate his method. He completed it with over 4 million bricks as a feat of engineering and mathematics that defied learning at the time (he left behind no plans or drawings).

The combination of aesthetic creators, enlightened activators, and aspirational culture labs (in Florence and other European cities) eventually led to the formation of schools, orphanages, the printing press, and the Sistine Chapel.

Our current renaissance is being activated in a similar way, only today aesthetic creators are not just contemporary Brunelleschi geniuses but grassroots creators everywhere—and activators are not just the Medicis of today but grassroots activators all over.

Actually, it is hard to imagine broadly shared renewal coming about in any other way than through a grassroots renaissance. Back in the 1400s, the future of most Europeans looked essentially like their past (notwithstanding whatever Gutenberg, da Vinci, and Brunelleschi might do). Institutions like the Catholic Church did face threats, and perspectives that held society together were set to evolve. But how people ate, managed their health, traveled from one village to the next, and the general nature of human life would fundamentally remain the same for most people for a very long time. Our situation is different. We see the human condition changing fast in the new things that enter our lives, from augmented reality to robots and artificial intelligence, and in many indirect ways, as reflected in fast-evolving social mores and the new nature of the workplace.

What shapes renaissance living is not so much the nature of what will be invented today, but what we inherit from yesterday, which in our case means a stream of innovation coming at us from the first two ways of creating things. For the future to be one we all want, however we manage to resolve planetary sustainability challenges ranging from where we get our food to how we manage climate change, we need to learn how to thrive in fast-changing circumstances as pioneers who continue to learn and discover.

Grassroots activation is happening all over today. More than the giving of money to creators, grassroots activators meaningfully engage. Engagement is what parents, friends, and lovers do when they give attention, time, and material resources to their children, friends, and partners to help them realize their personal aspirations. Helping our friends or our children realize their dreams is in a way sharing these dreams, while in another way, it is sharing the experiences that promote their dreams.

During his middle-school years, my son Raphael, a keen observer, had a dream to make movies. My wife and I bought him cameras and helped him think through his various film projects. Sometimes we even acted in them. When he turned sixteen, Raphael became interested in how the human body works. He had some problems with his knees from growing too fast and suddenly had a passion to create things that would help people with physical disabilities. It was easy to support Raphael as his dreams evolved. As parents, seeing your child discover a passion is naturally thrilling.

Grassroots engagement works like this.

Angel investing can be a consequence of grassroots engagement. With roots in the patronage of Broadway theatrical productions, angel investing has steadily grown alongside the rise of the GCM, to the point that today as much money is invested into the US economy by angel investing as by so-called venture investing, which follows more predictable investment guidelines aimed at strong profits over foreseeable time horizons. Angel investors range today from the one-dollar-a-month contributors investing via crowdsourced funding websites to those who place bets of time and money on students, entrepreneurs, artists, chefs, designers, scientists, engineers, and volunteers.

As Brian Cohen writes in his popular book *What Every Angel Investor Wants You to Know*, angel investing has been rising these last years for two basic reasons: the cost and time required for creating things have continued to drop, while the availability of resources has continued to rise. Cohen points out that among the resources that angel investors commit, none is more important to success than what he calls "intimacy." According to Cohen, in successful angel investing, creators and activators "need to stay close, and not just during the courtship phase of the relationship. The mentorship that founders so desire and the mentorship angels are willing to offer really defines a mutually beneficial relationship."

A form of angel investing with almost no intimacy at all is crowdfunding, which dates back to altruistic campaigns like the nineteenth-century construction of the Statue of Liberty's pedestal. The first crowdfunding website ArtistShare appeared in 2003, followed by Indiegogo in 2008, and Kickstarter in 2009. But creators like those whose stories appear in the last chapters rarely turn to it, and if they do, they soon move from crowdfunding to other forms of support that provide the intimacy to which Cohen refers.

Angel investing without intimate engagement is like the energy of a spark without the oxygen that makes it grow. That crowdfunding is such an important source of financing of the GCM (by 2015, total crowdfunding investment around the world amounted to around $34 billion, compared to around $141 billion global venture capital investment in the same year) sadly indicates that much of the GCM has little to do with long-range transformative creation. Crowd-sourced funding can obviously lead to created things that matter. A good example is the Statue of Liberty itself, conceived in the immediate aftermath of the US and French civil wars. Money is generally useful for getting something made. But barring extraordinary circumstances like those of Lady Liberty, grassroots crowdfunding has less likelihood of promoting the learning of aesthetic dimensions than more engaged grassroots activation, as I describe next.

How Transient Creation Becomes Aspirational

Terry McGuire had a Dartmouth engineering degree, a Harvard MBA, and several years of startup experience. The year after launching Polaris Venture Partners in 1996, he came upon my article in *Science* on the topic of inhaled insulin and wanted to discuss with me the possibility of leaving academia to start a company. I was flattered by McGuire's curiosity and a little nervous about it too. I was satisfied in academia and saw my future there, so my mind was divided. It excited me to imagine an idea spinning out of the university and "coming to life." But the notion that I would leave the security of my academic job and risk my career by tying it directly to the success or failure of one of my ideas seemed dumb.

McGuire came to see me at Penn State University, where I taught at the time. He was relaxed, smart, kind, and reassuring—someone I would enjoy getting to know. He invited me for an ice cream at the campus

Creamery. We bought a couple cones and sat down outside at a picnic table.

As soon as we were settled, McGuire asked, "So what's the business model?" The question was rhetorical. He knew I had no clue. Already I was off balance. If McGuire wanted to lure me away, he had started down a curious path.

"Normally, you have a drug, or you have a device. You don't have either the first or the second." He paused and let the depressing reality sink in for a little bit. Then he said, "You have a particle. A cool porous particle that nobody else has. Maybe the business is like this: You're in the business of sending people into space. What do you need? You need good astronauts—those are the drugs. You need the spaceship—this is the drug delivery device. But you also need rockets. You've just invented the best rockets there are."

"Insulin is just one astronaut," I replied. "We can send others. And the spaceship doesn't matter. We can make almost any spaceship work."

"Can you make enough of those rockets?" He looked at me expectantly.

I waved at the door to the Creamery. "See this place? They make lots of milk. I bet I could get access to the equipment they use to make dried milk and see if I can make porous drug particles—by the kilogram."

I was into it. I wasn't sure whether I wanted to leave academia and start a company, but McGuire had activated my interest by pointing to a creative process that was familiar to me and an idea that I cared about. Months later, I'd figured out how to make my particles on a large scale with the help of my students, and we started the company.

McGuire became a great advisor. Eventually, he helped me learn how to think through a startup pharmaceutical business, how to win partnerships, and how to make an investment pitch.

Following the sale of our company and the burst of the tech bubble, euphoria died in the biotech startup world. Industry focus fixed on blockbuster products. Meanwhile, health care researchers grew sensitive—partly as a result of the unprecedented global health care commitment of the Bill & Melinda Gates Foundation, and partly out of media attention that shed light on growing inequities driven by late 90s wealth gains—to the infectious disease health care burden that grew in places not yet served by commercial markets. In this global health care context, I now wondered if the path McGuire had led me along might hint at one I could take to effect change with no immediate commercial promise. Tony Hickey was then a professor of pharmacy at the University of North Carolina. He did aerosol drug delivery science, like me. At an international conference I attended, he gave a talk on inhaled antibiotics for treatment of tuberculosis in the developing world. He was trying to solve a health care problem that had no economic incentive. The work mattered to global human health, but it did not much matter in the context of the normal commercial innovation model. He was innovating but following a different path. By his thoughtfulness and generosity, Hickey activated a dream that had been percolating in my mind.

When Bill and Melinda Gates decided to launch their Grand Challenge program in Global Health, calling for proposals around the world to address thirteen big health care needs, I applied with an idea to bring the technology I'd developed for inhaled insulin to the developing world. The technology had been bought by Alkermes three years before. I asked the Alkermes CEO if we might carve out the technology for use in the developing world, where our patents had no applicability. This turned out to be complicated, probably impossible, as it raised the chance that a far lower-priced competitive product might emerge elsewhere in the world and undercut the developed world market. I pressed ahead anyway, motivated by Hickey's example. Around that time, I encouraged a group of undergraduate students at Harvard to look into forming a nonprofit should we have a technology—one I now needed to invent—to bring TB drugs and vaccines to the developing world. My students found a way to travel to South Africa and met a scientist there named Bernard Fourie, a world-renowned TB scientist and a descendant of one of the original Huguenot families in South Africa. Eventually, I went to South Africa and met Fourie. A few years later, he would head up our nonprofit in Pretoria, Medicine in Need, or MEND.

While working on the science behind MEND in those years, I met a TB scientist, Anne Goldfield, who introduced me to the *Time* photographer James Nachtwey. Anne described how Nachtwey's visit to her TB clinic in Cambodia had changed the way she ran the clinic. Seeing her patients through the lens of one of the world's leading war photographers helped her "become a better doctor." Nachtwey and Goldfield showed that the future of health care would rely on more than science and technology; it was about arts and the humanities as well. When I opened Le Lab in Paris, a first exhibition involved the photography of James Nachtwey, which documented scenes of doctor-patient contact preceding death in the clinical work of Anne Goldfield.

A decade later, what had begun as a special particle to deliver insulin to the lungs became a platform for digital health that used signals of scent to act on olfactory receptors, modulating neurological states and metabolism. This created thing emerged out of constant and diverse grassroots engagement.

Most sustained pioneering experience advances this way, step by collaborative step. When we try to build from the transient experiences of the GCM long-term creative experiences to effect change, we need the participation of others. When innovating while following the first two ways of creating, we can point to concrete value that we produce—for learning, for commerce, for culture, or for society. When following the third way, we are constantly searching for what is intriguing, provocative, or in a sense beautiful. This requires deep focus and dedication, making us especially needy of support. Activators provide this support, as did the patrons of the Italian Renaissance.

Today activators can be anyone. I have many in my own creative life, perhaps none as intimately engaged as Jim Buzzitta, a gregarious medical doctor and health care leader in western Michigan, grandson of Italian immigrants.

Buzzitta had been one of my dad's students back when my dad had taught organic chemistry at a community college. I had met him a few times when I was twelve or thirteen, and then we'd lost track of each other. In 2013 a media crisis related to the US launch of my first air food product (Senator Chuck Schumer had demanded that the FDA look into the product at a press conference—we met all the demands of the FDA in the end, but the public path we took was stressful and new to me) broke just as my dad acquired a septic infection. With a lobe of his lung removed, a hole in his heart, and some signs of spreading lung cancer, my dad entered ICU care. After a Skype interview from my Paris kitchen with ABC News, I jumped on a plane from Paris to Miami, rented a car, and drove up the coast to Flagler Hospital in St. Augustine. I made it to my dad's bed for a night vigil, as his body pulled through the infection scare, and the next day, as I groggily carried on, Buzzitta flew in from Michigan.

We spent several hours together at my father's bedside and then went for a walk around the hospital. I had come to Florida to help manage my dad's medical treatment and my mom's anxiety and fears, while I struggled with my own emotions. Buzzitta strengthened me with insights from the years he had known my father as a student and from his own very full life. As my dad's situation started to improve, Buzzitta proposed coming to Paris with his son Jack. They came, and we began to work together on what had become by then the seed of my digital health platform.

My businesses often kept me in the United States for prolonged periods of time, away from my family in

France. I was alone and lonely. I now had a restaurant next to MIT where I often sat and worked, surrounded by strangers, friends, and occasionally people I'd known twenty-five years before, when I'd been a student. The inventions I'd exhibited in Paris as works of culture, chocolate that entered your mouth through the air or vanilla ice cream bites surrounded by edible packaging of chocolate and caramel, now sold in the restaurant and stores across New England. As these things I'd created began to surround me, I felt like an altered person. My wife led a lab for renal physiology in the center of Paris. She had signed on for a five-year stint. At the time she'd made the commitment, we had imagined staying in Paris indefinitely, perhaps the rest of our lives, shuttling back and forth to the United States. But as my inventions commercialized, all of the resources came from the other side of the Atlantic—and Switzerland, by way of an angel investor friend, Bernard Sabrier. Pioneering these new things toward actual health care impact (and not losing my shirt in the process) meant I needed to be close to my activator network in the United States. This put Aurelie in a situation where she had four years left on her contract and a husband thousands of miles away. Her initial thought, in the spring of 2015, several months after I'd opened Le Lab in Cambridge, was that she would move to Boston after our youngest finished high school, at the end of her lab direction mandate. She had sacrificed her career during the first years of our family, working part-time, and finally landed this incredible opportunity in Paris. If she left, she would probably never find such an opportunity again. It was unfair for her to leave early, while I doubted I could survive a transition of several years cut off from my family. I was frustrated and confused. It occurred to me that in selling a dream, I may have unwittingly sold my personal life. In the spring of 2015, alone on my birthday while floating on the boat in Boston Harbor where I'd mostly lived during those years when our home base was in Paris, I grew sad, a rarity for me. Living away from my three boys during their critical high school years, I felt a guilt and sorrow that proved too difficult to bear on my own. It was Jim Buzzitta I called that day. I remembered his familiar words—"Always family" ("Sempre famiglia"). It was an expression that depressed me at the time. He had been through divorce and told me I would survive it if that's where we ended up. Nothing was perfect, but it was something we could do to make our lives better than they were. The last two decades mingled together in my head. I couldn't psychologically separate my personal and professional lives. Over the months that followed, I called Buzzitta every week. My eldest son, Jerome, moved to Boston and started his junior year in high school in the fall of 2015. By then, we planned for Aurelie and our other two boys to join us the next summer. Neither Aurelie nor I were sure we would. We struggled month by month. Buzzitta took my calls any time of the day or night, and he sometimes came to see me, while the experience of living with Jerome began to change me. Most nights we ate together. My son and I read together in the living room, watched American football, and became closer than we'd ever been before (or after). My life came back into balance. I spoke to Jerome honestly about my feelings, and he spoke honestly to me about his own. Once in those months, we got on a plane and flew to see Buzzitta, where Jerome spent a day with Buzzitta's son Jack. Buzzitta's friendship bought me the time I needed to discover what mattered to me, and the time Aurelie needed to discover the same thing for herself. We eventually rediscovered each other while getting our youngest kids to the United States and Jerome into university.

Buzzitta activated me as a creator, but he also partnered with me in my life, as a son, a father, and a husband. At the end of his text messages, he often wrote, "Love you."

Aesthetic creation emanates from raw individual experience and can seem hard to justify, let alone understand. The challenge of its process and the value of its outcome can each get easily overlooked. Activators change this. They care long before the merits of the project argue they should.

Activators give creators courage to pursue their dreams. As they do, transient culture labs become aspirational. They develop business models that help creators guide investment toward the fulfilling of dreams, in the sense that El Bulli helped Ferran Adrià, Orion helped Richard Garriott, and ART helped Diane Paulus. They deliver immediate cultural value while permitting creators to learn within a Creator's Cycle that can repeat indefinitely.

Why We Create Aspirational Labs

Aspirational culture labs incubate the third way of creating. They facilitate ideation, experimentation, and exhibition, as do transient culture labs, and they do something even more important: they value what matters to the creator and help connect this thing that matters to everyone else.

The Florence of Brunelleschi again provides a helpful example. Young Brunelleschi had a passion to revive the architectural prowess of ancient Rome after having spent several years in Rome with his friend Donatello studying imperial ruins. Back in his native Florence, he had won a formal competition issued by the Arte della Lana to finish the dome on the Santa Maria del Fiore, whose nature and dimension had not been imagined since the days of ancient Rome. What mattered to Brunelleschi mattered to Florence. Then, over decades he received commissions and support to iteratively arrive at the expression of his passion in the creation of the Duomo, which finally mattered to the entire world. Florence gave the oxygen Brunelleschi's candle needed.

Aspirational culture labs encourage the "translation" or movement of created things from the lab into society. To make this possible, they provide some organizational structure, funding, and connections to commercial and cultural centers outside the labs. Through frequent public engagement, they help us realize dreams. They can be any organization—company, nonprofit, academic lab, even government—and have an unmistakable altruism.

The first government of the United States was a kind of aspirational culture lab. It was led by creators who had developed a dream of democratic government from the Declaration of Independence through the revolutionary victory over the British Crown to the Treaty of Paris. They honed their idea from old models and contemporary theories of governance and ultimately through the experience of running the first colonies. They pitched the idea as a kind of work of art in the US Constitution. Electing George Washington as the first president hardly guaranteed a long-standing successful democracy. The shape of American Democracy, the ultimate created thing that mattered, took years of activation, beyond the end of Washington's presidency. From the Federalist Papers of Hamilton to the Gettysburg Address of Lincoln and long after, the United States government survived because it evolved.

Aspirational culture labs are like this. Unlike most companies, nonprofits, and governmental organizations, they retain a freedom to restate their mission to guarantee survival of the dream.

Le Laboratoire is a particular kind of aspirational culture lab, open to the public. It came to life in 2007, also soon after the first tweet, and simultaneously with two other public culture labs in Europe. All three opened with an intention to explore, with the public, science through art and design. In London, Ken Arnold opened the Wellcome Collection in London, where connections between contemporary art and medical science are examined, stirring public debate. In Dublin, Michael John Gorman opened the Science Gallery, where works of art and design are used to explore frontiers of research, such as nanotechnology and tissue engineering. Since opening, Science Gallery and Wellcome Collection have grown and expanded to new cities around the world, and Le Lab moved to the United States.

I modeled Le Laboratoire on the traditional science lab, where discoveries are made. From these discoveries, technologies, products, and works of culture eventually move out into the places where they can bring enduring value. In science labs, researchers get ideas and funding from a specialized community of activators (peers, corporations, foundations, government). Together, researchers and activators explore, publish among peers, patent ideas that seem practical, and spin these patented ideas out into companies that develop them into technologies and businesses. We wouldn't recognize our world without this model. But the science lab model works with a well-defined set of problems and opportunities. It does not work to explore a really broad question like, why do people do terrible things to each other? (A science lab might break this question up into its medical, social, and political components and produce, among other things, drugs, technologies, and policies.) Nor does the traditional lab deeply involve the public in the exploration process. In creating Le Lab, I was interested in a lab that *did* involve the public, not as audience but as participant; that *did* approach large ambiguous questions; and that followed the science lab model by producing things of long-range public value.

Le Lab became my Florence.

As creators, we all need a Florence. Florence is in fact the thing most wanting inside the GCM today. We are not lacking creators, resources, or goodwill to connect the two. What we mostly lack is the oxygen, or culture, which can give people everywhere the confidence that what matters to them can matter to many others, and the guidance that points their aspirations in fruitful directions.

From antiquity, cities, regions, and nations have long brought to individuals a culture of change. Today in the United States, for example, we have large-scale aspirational culture labs in Silicon Valley (for information technology), Boston (for biotechnology), Hollywood (for cinematic film), and New York City (for Broadway theater). Obviously, all these powerful incubators of creative change are valuable, but in the era of the GCM, we especially need grassroots aspirational culture labs.

We need more Judah Folkman mentors in frontier science labs, more Paul Sills mentors in experimental theaters, and more Uncle Georges in living rooms.

Silicon Valley changed the world, as Florence once changed the world, and ancient Persia long before then. But our challenges tomorrow, like our opportunities, are increasingly local, systemic—personal. They reflect the grassroots interchange between our biology and the fast-changing circumstances we sense around us. We get a hint of this grassroots interchange in the incessant buzz of billions of tweets.

CHAPTER 8

Culture, Change, and Hope

ONE AFTERNOON DURING THE HOT SUMMER OF 2017, I visited the United Nations Secretary General's office in New York City to talk about the global implications of scientific research. My brother-in-law, Marc Jacquand, had just been made an advisor in the office of the Secretary-General António Guterres and had invited me, with the director of Paul Allen's Frontiers Group, Tom Skalak, to meet the Assistant Secretary-General Fabrizio Hochschild and his director of strategic planning, Ayaka Suzuki. Hochschild, a savvy veteran of international politics and peacekeeping operations with deep experience in some of the most troubled regions of the world, began our conversation by describing the atmosphere reigning at the United Nations.

Overall, he said, things looked bleak. The world was in crisis—socially, politically, ecologically—and the UN remained sunk in conflicts in Syria, Yemen, and even in forgotten Cyprus. The seventeen UN sustainable development goals—to end poverty and hunger, improve well-being, bring education to everyone, gender equality everywhere, among other lofty aims—seemed impossibly remote. Whatever the UN did with its limited resources, humanity floundered, while reports of gene editing, artificial intelligence, and the new generation of robotics made it seem as if science was about to edit out humanity altogether. What was to be done about the future?

Hochschild's question was rhetorical. Arms crossed, I waited for him to continue. Skalak, however, jumped in.

Gene editing and artificial intelligence were frontier explorations that the Paul Allen Foundation funded. Skalak admitted that some might see these science programs as threats, but in his view, they weren't at all. "What happens at frontiers of human experience today is fascinating," Skalak went on. "We live at a time of unprecedented intellectual freedom and discovery. But you can't discover the future fearing the harm exploration might bring. There's no guarantee that any frontier will produce a useful discovery, but if the risk of discovery extinguishes the drive to pioneer, we're already doomed." After a pause, Skalak added, as if by afterthought, "Obviously, we all need water, and we all need food. But without hope, we have nothing."

We had come to the UN in the context of the first World Frontiers Forum, which was to take place later that fall. The idea of the forum had come out of a conversation I'd had with Bob Langer and Dennis Ausiello—

emeritus chief of medicine at Massachusetts General Hospital and a pioneer of personal medicine at Harvard Medical School. Le Lab would soon celebrate its tenth anniversary. We thought to invite leaders from around the world to experience fifteen frontiers of tomorrow, all at once. It is hard to understand the future of food, for example, and not the future of biology, transportation, energy, and climate change. In a sense, you can't imagine the future without every pioneer in the conversation.

Until this moment, I had understood the World Frontiers Forum to be about dreams of a sustainable tomorrow. Listening to Skalak, I decided it was even more profoundly about the hope we can offer today. Dreams could take decades to make a difference in the world—as had Danny Hillis's dream of pinch-to-zoom or as might Neri Oxman's dreams of material ecology. Without hope, they had no chance. You needed to be patient with dreams, while hope was immediate and unifying.

At one point, Suzuki asked, "Who should be in the room to discuss the future? We used to believe representatives of the earth's nations could speak on behalf of the peoples of the earth. This doesn't make much sense anymore. National governments are not driving change. People everywhere are."

This idea that without "everyone in the room" you couldn't have a unified conversation about the future provoked me. Even if we could get everyone together, what would we possibly discuss? Robotics? Synthetic biology? I found it hard to envision a relevant topic of conversation that would lead to anything near a consensus. More than information and logic kept people apart. Experiences and beliefs did. An electronic avalanche of information had diminished the power of information to bridge the gaps between us. With all the chasms, it was hard to chart a future together. What drove pioneering creators—hope in an *uncharted* future—came from rare and moving experiences at frontiers, like the discovery of a tranquil starry night after a wicked storm at sea. The thrill of pioneering experience was tied to its surprise. Hope in "not knowing" came out of years of learning and honing of the aesthetic dimensions of creating. Somehow in this room we needed to share *this*.

Hope in an uncertain future finally counted more than the created things that shaped it.

Many things deliver hope in our daily lives. Some are near us and easy to act on, like turning on the television or going out with friends. Others may be far away and take advance planning, like the dream of getting a college degree. When conditions change on us, or for whatever reason we lose grounding for hope in the future, life can feel "uncharted" or frontier-like. When this happens, we sometimes latch onto sources of hope that are reassuringly nearby and forget about this future we don't understand. Otherwise, we take the initiative, expressing new experience (asking great questions, writing songs, starting a business) in careful ways that draw the attention of others, make them stop and notice us, wake them up. Sharing pioneering experience, we start a kind of conversation about the future. It is this conversation we need to have today.

Suzuki's metaphorical "room" for figuring out the

future is not a single room at all but a collection of rooms. More literally, these rooms are the surprising experiences we share that move us in similar ways. Many of these have little to do with human creativity and include snowy blizzards, market crashes, and human wars. Other experiences—as in reading *Uncle Tom's Cabin* by Harriet Beecher Stowe, or in listening to certain national anthems like the *Marseillaise*—are the outcome of creative process itself. They are the aesthetic works generated by every iteration of the Creator's Cycle that propel a cultural conversation about the creating of tomorrow.

Hope and Cultural Experience

Massimo Bottura is arguably the world's leading practicing chef. His restaurant Osteria Francescana in Modena, Italy, has received first or second ranking on the San Pellegrino list of the top fifty restaurants in the world for the last five years. Modena is a small town in Italy's Parma region and is famous for its automobiles (Ferrari, Lamborghini, Maserati, among others) and local food (balsamic vinegar comes from Modena, while Parmigiano-Reggiano comes from the Modena region that gives the cheese its name).

In 1992 Bottura, who was born in Modena and had recently brought his American wife, Lara, back from New York City to live with him there, made an experimental dish entirely of Parmesan cheese for the renowned French chef Alain Ducasse. Bottura's idea was an experience of Parmesan with three unique textures—crispy, creamy, and airy. The next year, around the time he first introduced the dish into his restaurant, he received a guest named Umberto Panini, who had just opened a factory for producing Parmesan cheese. Panini explained to Bottura that three textures of Parmesan did not only mean three temperatures; they actually implied three very different ages of Parmesan cheese (a fifty-month-old Parmesan cheese has a completely different texture and taste than a twelve- or forty-month-old Parmesan). Bottura began to experiment with different ages of cheese. His dish evolved over the next two decades, from three to four to five textures.

In 2011 I had the chance to dine at Osteria Francescana. My family had spent the day with Lara and "Max" and their kids, among other things discussing what we might do to further an experiment we had started the year before making "flavor clouds" using a special carafe I'd invented at Le Lab with food designer Marc Bretillot. Our carafe allowed you (by the action of ultrasound generators in the base of the carafe) to form a cloud out of almost any liquid—from a martini to a tomato soup—that you sipped into your mouth. Naturally, Bottura had started to use it to make clouds of Parmesan. That evening, while Lara generously took care of all the kids, Aurelie and I walked to Osteria Francescana, where we met Bottura, who spoke to us for about thirty seconds (the day's collaboration was done; he was visibly preoccupied), and we sat down for dinner.

The famous cheese dish came out soon enough. It looked like yellowish brushstrokes on a white canvas. We started by delicately placing in our mouths a fifty-month-old Parmesan blended foam with a smidgeon of hot demi soufflé (the twenty-four-month-old Parmesan). Then we mixed in our spoons brittle fragments of the forty-month-old cheese, which had been prepared as a thin galette, with the warm creamy sauce of thirty-month-old Parmesan and the thirty-six-month-old variety, which had been cleverly siphoned onto our plates as chilled foam. The dish blew away the evening with a single ingredient.

Massimo Bottura's aesthetic expression of deeply personal experience growing up in the unique small town of Modena was neither a lesson in the aging complexities of Parmesan cheese nor was it very filling. It was a sensory experience that altered our perception of food. And not just us! This same year "Five Ages of Parmesan" had been named Italian Dish of the Decade.

Shared aesthetic experience is often a consequence of human creativity that has little to do with conscious artistic process. I remember watching the first landing on the moon in the summer of 1969, when I was eight years old. I sat with my parents and sisters in our living room, knees curled up beneath me, watching along with everyone else as Neil Armstrong softly placed his feet on the surface of the moon. I was a little confused as I stared at the television screen. The Apollo 11 lunar landing was not nearly as visually clear (nor as exciting) as the Batman rerun television series I liked to watch in those days. I was obviously very young, and antsy, and I'm sure I was anxious to go outside and play. However imperfectly, the medium of television gave me an experience of the first human step on the moon. The emotive power of that step grew on me as time passed. For years in grade school and on through middle school when kids talked about what they wanted to be when they grew up, many said "astronaut" with a nod—literal or not—to that day we all watched Armstrong walking on the moon. My teachers liked to show pictures of the Apollo 11 capsule when they talked about the Russian-American Cold War or when they tried to make us care about the physics of the solar system. My generation grew up looped together by shared memories like this one, and such memories made us interpret events in similar ways, as when we all celebrated as new and revolutionary Michael Jackson's moonwalk (the dance step had actually been around for decades). It wasn't that we needed to see another moon landing, any more than I needed to enjoy again the Five Ages of Parmesan to remain changed by it—our passions turned to things more terrestrial. But we had shared an experience that showed us almost anything was possible if humanity put its mind to it, and that ambition could lead you even further than you wanted to go.

Fresh new cultural experiences like these unite people around common beliefs, desires, and fears, a reality we implicitly acknowledge when we make references to, for instance, "the Space Race era," or, in other contexts, to "the impressionist era" or "the postwar era." Our era titles refer to experiences of groups of people that provoke, upset, elate, confuse, and in any case, surprise in similar ways. We pay more attention to the world around us after having these experiences than before we did. We are more prone to discover what comes next and, however soberly, more hopeful of our ability to navigate an uncertain future.

Such unifying experience is a high calling of aesthetic creation—in the arts as in the sciences, even in a restaurant.

To shine a light on the relevance to the future of these unifying experiences, we came up with the idea of the World Frontiers Forum.

The World Frontiers Forum

We held the first Forum in the fall of 2017 at Le Lab, Café ArtScience, and Harvard Business School. Dialogue across cultural and political boundaries seemed to disappear in the United States over the eighteen months that preceded the first Forum. In pioneering circles, there was thirst for a place where we could actually get along, find a common language, and share hopeful ideas of tomorrow. The World Frontiers Forum is about creating the future of not five but thirty years from now. The Forum gathers leaders from around the world in an interdisciplinary pioneering spirit and aims to propagate this spirit as a compelling public experience through a new work of art and science somewhere in the world.

At the inaugural Forum, former MIT president and biologist Susan Hockfield talked about what she called convergence. Fields like biology and engineering are fusing as they attempt to meet the basic survival needs of over nine billion people on the planet. She spoke about molecular structures called water channels that appear in cell membranes and about her visit to a Copenhagen-based company called Aquaporin that is engineering these water channels to filter water on what may one day be a massive scale and at an affordable cost. Temple Grandin, charismatic advocate for the autistic, spoke of her experiences growing up as an autistic learner and the importance for all of us to level the playing field for those who see the world in original ways. Later the same day, Cato Laurencin, the medical doctor and pioneer of regenerative medicine from the University of Connecticut, shared a vision of the future where the military surgical handsaw disappeared from the field of battle (he showed a pair of images of handsaws that had hardly changed in the 130 years between America's Civil War and the Iraq War). With regenerative medicine and engineering, Laurencin foresaw the day when severe wounds might heal themselves and severed human limbs grow back to their original forms. The former New York City Ballet dancer Wendy Whelan performed an original dance with choreographer Brian Brooks. Joining us from the United Nations, Fabrizio Hochschild discussed the challenges of cybersecurity, growing global hunger, and a general sense of malaise about the future. If organizations like the United Nations could experience frontiers as places of possibility, and scientists become more aware of the reality of the most unfortunate, the future might appear less frightening. Sam Kass, a former chef in the Obama White House and food entrepreneur, spoke about the dramatic changes global warming will bring to how we eat. As he spoke, everyone enjoyed and mulled over what Kass called the Last Supper, imagined with Café ArtScience chef Carolina Curtin, where each course presented natural foods, like oysters, chocolate, and wine, which will be hard to access when the planet heats up by two degrees.

The Forum took place in the midst of an exhibition of Danny Hillis's Clock of the Long Now, which the curator, Hans Ulrich Obrist of the Serpentine Gallery in London, called "one of the most important art projects today." The project reflects the kind of engaging and provocative public artwork imagined as outcome of the Frontier Art Prize.

We awarded the Frontier Art Prize to Doug Aitken two weeks later at the Picasso Museum in Paris, on the opening night of the FIAC International Contemporary Art Fair. In a conference before the award dinner, where Obrist and other members of the jury and the art world gathered, Aitken spoke about his underwater pavilion installation off Catalina Island. Joining Aitken was Fanny Douvere, who leads UNESCO's world maritime heritage sites. Douvere warned of the "numbing effect" on people of the growing bank of alarming scientific climate change data, while the screening of a magical video of Aitken's underwater pavilions riveted the audience and drew a few smiles when a seal showed up and began to explore one of the pavilions. Douvere wondered if art might provide a route to help better tell the scientific story of the impacts of climate change on the ocean.

Aitken said art gave people fresh experiences that were powerful enough to change their minds and lives. In Aitken's view, thirty years of digital immersion and an ever-ramping flow of data had bred a contemporary thirst for material experience that came no faster than our senses allowed, the kind of surprising, invigorating, and elusive sensory experience artists have always brought to the human condition.

Generosity & Cultural Experience

Creating new aesthetic experience is doubly generous. Not only does it give value to others; it produces value that never existed before.

Discovering and sharing value can take decades of trial and error. It usually involves sharing things (manuscripts, prototypes, imperfect compositions) that are of less value than they promise to ultimately be. Falling short of their promise, creators run the risk of never producing enduring value at all. Established writers, scientists, composers, and other seasoned creators look for mentors and activators of all kinds who will pay attention to what is incomplete. This is what happened when Judah Folkman attended to Bob Langer over the couple years it took Langer to create his bio-eroding polymeric drug particles, or when Paul Sills attended to Diane Paulus over the years it took Paulus to shape her passion for improvisational theater. By the gift of their care, mentors and activators help a created work eventually become a thing that others want, a gift to the world.

In their book *The Paradox of Generosity*, Christian Smith and Hilary Davidson show that generosity, selflessly giving to others beneficial things, paradoxically increases well-being in the lives of the generous. In their national study of generous habits in the United States, Smith and Davidson focused on a few basic forms of giving, including donations to others of 10 percent or more personal income, volunteering, and consistently helping neighbors and friends. They looked into how generosity impacted happiness, bodily health, purpose in living, avoidance of depression, and interest in personal growth. Those who gave regularly generally felt better about themselves in most or all of these ways than those who did not.

Altruism excites the reward cycle of the brain. Beyond this, psychological studies have shown it can increase our personal agency and give us positive social roles, reduced self-absorption, perceived abundance, enlarged social networks, and more opportunities for learning and physical activity. All these benefits come to us when we collaborate with others to achieve what is broadly valuable. They deepen and endure over the long process of creative engagement that leads to creating things that matter.

But do the things we give to others (whether money, personal time, or a created work) actually bring to others the benefits we think they do? Actually, the entire arc of the expressive life I have described in this book depends on our success in benefiting those we meet—collaborators, activators, first publics—along the way. Creating can be thrilling, but its sustainability is tied to its true generosity. We may dream that a created thing will one day eliminate poverty and disease, but if our path to realizing the dream brings no benefit at all, does nothing to change minds and lives, we will never get there.

The Expressive Path of Creating

My interest in the future of food grew out of my work in Africa with inhaled drugs and vaccines for tuberculosis. As relative affluence came to Africa, the kinds of health issues that plagued the United States, primarily chronic diseases related to metabolic and neurological dysfunction, grew alongside the infectious disease crisis. Food waste, including plastic refuse, started piling up outside major urban areas where I traveled, like Johannesburg and Cape Town. I began to believe that some of the innovations that medical science and technology had brought to drug and vaccine delivery might be brought to food without the tremendous costs of development and the regulatory hurdles that kept many of our medical innovations from practically changing health care on the ground.

In the first year of Le Laboratoire, we did an experiment with French Chef Thierry Marx and the Parisian scientist Jérôme Babette involving a brand-new technique of spherification, a food encapsulation process made famous by Ferran Adrià. I wondered if it might be possible to breathe food into our mouths, to ingest it using what we'd learned while developing inhaled insulin. Over several months, my students and I invented Le Whif, a way of getting chocolate into your mouth by inhaling it. With Le Whif, we could deliver a flavor experience of chocolate with essentially no calories; later we showed how we could eliminate pills and in some cases improve efficacy by delivering nutrition in the same way, like Vitamin C or B12 (about 40 percent of people have an aversion to taking pills, meaning they often fail to take nutrients and medicines that they should). I collaborated with French food designer Marc Bretillot on our "air food" carafe, which we called Le Whaf. By making clouds of flavor, ranging from coffee to cotton candy, the carafe showed more clearly the sensory possibilities of air food. A couple years after creating Le Whaf, we held a Flavor Cloud Weekend in Paris, where I first met Massimo Bottura, together with other top chefs including Ben Shewry of Australia's Attica and Homaru Cantu, then owner of Chicago's Moto. Bottura presented a cloud of orange over duck reduction, while Shewry developed a cloud of sushi (you passed your glass through separate flavor clouds of ginger, soy sauce, tuna, and wasabi), and Cantu produced a cloud that involved his "miracle berry," which made sour tastes like lemon turn into sweet tastes, like lemonade.

Air food was a rare aesthetic experience, simple and ephemeral. I began to wonder around then about scent as a sensory signal, like light and sound, and imagined how it might carry emotive experience across digital networks.

I did another food design experiment at Le Lab with the French designer François Azambourg. Thinking back on the scenes of food-packaging refuse in Africa, I wondered if we could make packaged food as nature made fruit, with edible packaging. This led to an experimental restaurant, the precursor to Café ArtScience, which opened a few years later in Cambridge. The "FoodLab" in Paris was a first step I made to bring experimental food experience into a daily reality. I was curious to know whether such experimentation with people was actually sustainable. The longer you could hang out with the public, experimenting with people and learning together, the more likely it seemed that you would get to food rites of lasting value. In the Paris FoodLab, we served meals around novel food forms, which wrapped any food or drink in an edible skin or packaging, called WikiFoods. By 2013, WikiFoods had become a Stoneyfield brand that sold in Whole Foods stores around Boston. By 2015, the Perfectly Free brand of WikiFoods ice cream, which lacked the eight primary food allergens, had launched in Star Market and other stores in the Boston area. By 2016, sales expanded along the East Coast, and the company was now called Incredible Foods. Our ice cream bites and the Perfectly Free fruit bites that test launched in the spring of 2016 (and commercially in stores in the winter of 2018) had their own edible packaging, and you could just sell them like fruit and vegetables, without plastic. You would wash them as you did natural produce and not contribute to the global pandemic of plastic in our oceans and landfills.

In 2012 we opened an exhibition at Le Lab in Paris around the idea of texting the scent of coffee. This led to oNotes, a platform to produce scent signals with digital control, in the way we produced sounds signals with iTunes and a headset. Our "iTunes" of scent became an app called oNotes, and our "headset" of scent, a scent signal emitting device called Cyrano.

The olfactory nerve is the only sensory nerve that goes straight to the brain, particularly the hippocampus in the emotive region of the brain, which is the seat of long-term memory. Partly as a consequence, no sensation influences our physiology, emotions, and memory as deeply as our sense of smell. Integrating scent into digital experience establishes a special emotive connection to the digital environment, and we respond to this environment in ways that reflect and influence our physiological state.

The oNotes app provides digital control of scent sequences, called olfactory notes—or oNotes—like "songs" of scent. When we came out with Cyrano by oNotes, we imagined people would mostly use it in cars to calm and distract themselves on long commutes. This turned out to be wrong. In our first year of testing, we noticed that people used the platform more in stationary places than on the go, more in the afternoon than in the morning, and more during the week than on the weekend. It happened that people found digital scent a useful way to enhance performance. They used it mostly while "working" (at home, in the office, in a car) to shape well-being—picking up, calming down, helping psychologically escape (for instance "to a beach" with the smell of suntan lotion). We were discovering a new consumer behavior that solved a need, but it could still look odd—the platform got advertised with our flavor cloud carafe in the fall 2017 promotion for the futuristic movie *Blade Runner 2049*.

All this led to worker well-being testing of the oNotes platform, which we began at Siemens in Chicago over the course of 2017. While pursuing this work, Dennis Ausiello, cofounder of the World Frontiers Forum and Harvard Medical School professor, pushed me to think more deeply about the delivery of scent via the nose and mouth as a single sensory intervention with ramifications to metabolic health. Discoveries in biology and medicine, which increasingly point to the air as a medium of biological signaling similar to the blood that circulates in our veins, made it seem like we might use these new kinds of consumer experiences to curb flavor cravings and reverse food addictions.

Air brings scent molecules, or odorants, through nasal pathways and produces sensations of smell. When we exhale—and particularly when we hold aromatic things in our mouths, like a sandwich or soda—it again brings odorants through the nose. Our brains process inhaled scents as smell and exhaled scents as flavor. Flavor includes other sensations as well, like taste and touch (in the process of chewing). Flavor images, which form in our brains much like visual images when we see things, guide us to like and dislike foods. They also promote natural cravings, as when we crave an orange because we are thirsty and in need of vitamin C. These processes reflect natural metabolic regulation pathways that our bodies have acquired over human history in a relatively constant natural environment. Our modern synthetic environment, from the food we eat to the cities we build and the screens we use to communicate, so radically differs from what humanity has known that our bodies have become profoundly dysregulated. Today, a thirsty teenager can prefer a bottle of soda over a glass of water and develop unnatural food cravings and even harmful food addictions. The oNotes startup company that we'd spun out of Le Lab began to develop a research program with medical scientists in Boston and New York City aimed at a future of food experience. It was this future of eating that I had been intuitively exploring for several years, bringing together innovations in digital health care, information technology, and food.

One day, we might design a scent experience to help guide people to eat more wisely. As wearable digital health devices gather increasingly useful information about our bodies, it might be possible to use this information to produce sensory signals, including those of scent, to guide metabolic function, as our biology has evolved to guide metabolic function in the natural world. Food forms, including pure flavor experiences with various forms of air food, might be made for us based on our biology, a kind of "personalized food" that is custom-made, helping us while not harming the environment.

Many other scientists, designers, chefs, and entrepreneurs around the world today are on a path to the future of food that is experimental and shared, involving countless acts of creator and activator generosity. Pioneers like Pat Brown (former Stanford biologist and founder of Impossible Foods, who develops plant-based meat substitutes), Geoffrey Von Maltzahn (founder of Indigo, who develops microbiome-engineered crops), and René Redzepi (one of the leading chefs in the world and co-owner and chef of Noma in Copenhagen, who explores the culinary potential of insect-based food protein) guide people on experiential paths in restaurants, stores, and on farms, learning through and collaboratively shaping food behavior today.

Figure 16. Chef Massimo Bottura creating a cloud of orange during the Flavor Cloud Weekend at Le Laboratoire in 2011.

Figure 17. Le Laboratoire Paris (1st arrondissement).

Figure 18. Le Laboratoire Cambridge (Kendall Square).

Figure 19. Gallery entrance of Le Laboratoire Cambridge.

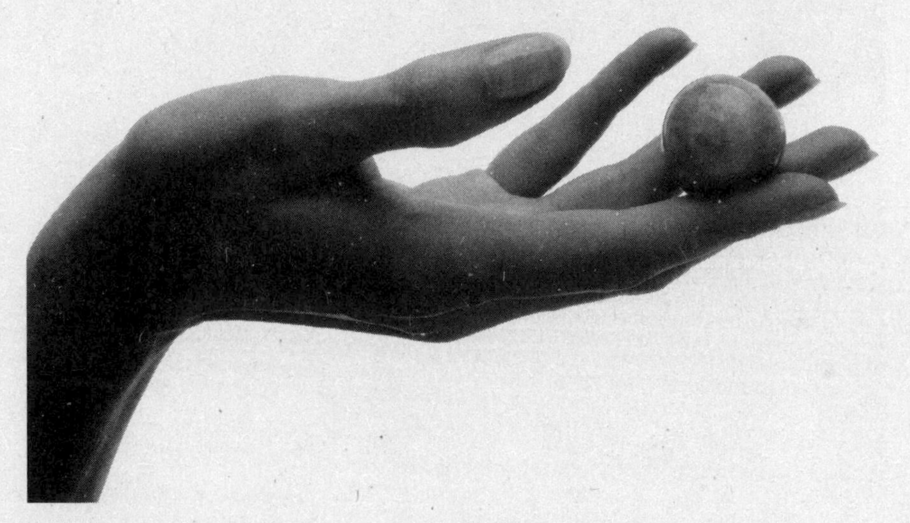

Figure 20. A WikiFood—or Incredible Food. The grape form contains vanilla yogurt inside an edible skin of kiwi.

Figure 21. The author with Le Whaf, the flavor carafe invented with the French designer Marc Bretillot.

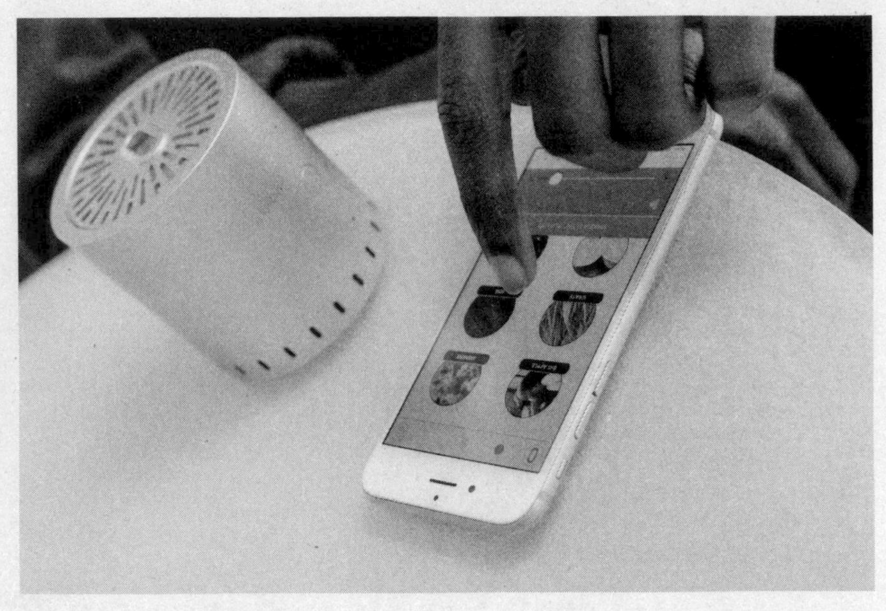

Figure 22. The digital scent player Cyrano with the oNotes app.

Maybe we will one day phase out plastic in food packaging, learn to sustain our food supply, and readapt human metabolism to our altered sensory environment. But to get there, the host of experiences we and many others bring to people while exploring this future of food will need to spread hope that it will happen. Inventing the future of food is impossible without building it on consumer rites people actually experience today, and enjoy, even as they learn to eat a little differently tomorrow.

Having Fun at Frontiers

During the first years of Le Lab, we ran an art and science "dream learning" program for high school kids in Boston and Paris. This continued the work of an educational foundation that my wife and I had set up after the sale of my first company, called the Cloud Foundation, and for a few years appeared in other cities, like London and Singapore. In Boston, each fall we had an after-school event where interested high school students from around the city could come check out the program.

One year we invited Brad Whitford, the Aerosmith guitarist, to speak to the kids. I introduced him on the stage of a large Loews cinema downtown next to the Boston Commons. The students applauded wildly, and I stepped back. Whitford had written out what he wanted to say on a set of three-by-five cards. He began to speak, microphone in his right hand, cards shaking in his left. Being the guitarist of one of America's pioneering rock bands was like being a kind of artist-scientist, he said. It

was also being an entrepreneur. A music band was, after all, a startup. He went on like this for a while, and then, all at once, he pitched his cards to the lectern, looked up at the kids without finishing his prepared speech, and shrugged. He said, "You know, I'm just having *fun*." Everyone laughed and went back to clapping, as if relieved to step out of this rationalization of what was, to them, the most perfectly natural path in life to want to pursue. The pioneering guitarist was having fun!

Whitford's speech was powerful. Beyond his words, there was the surprising experience of his awkwardness. He seemed dangerously sincere. His nervousness made me uncomfortable, and yet it was generous of him to be there with us, when he could easily have been playing his guitar before adoring thousands. Here he was, a rock star, standing in a Loews theater on a Thursday afternoon talking to high school students about why his life mattered. We didn't doubt it did, and so it came as a tremendous relief when he said he was having fun. Suddenly, we were all having fun with him!

We share the excitement and simple pleasure of pioneering creation too rarely. At frontiers, we do not necessarily balance in our brains the risk and opportunity of discovery and decide if the risk is worth it. Pioneering is hopeful because it is fun. It motivates us. Without textbook or teacher standing before us, we can figure out on our own how to survive and perhaps discover things that matter beyond us—to an Aerosmith degree.

Learning this way not only involves conversation and collaborative creation across disciplines and cultures. It

also means piercing the traditional boundaries of organizations and institutions that disconnect us.

As occurs today on many campuses around the country, at my university we have started to explore the possibilities of this experiential and aesthetic piercing of institutional boundaries. Inviting creators and activators of the GCM onto campus and sending students off campus into transient culture labs, our faculty hopes to help students learn to discover and have fun pioneering.

As I was finishing the writing of this book, a very special project began in a conversation with the dean of the Harvard Business School, Nitin Nohria, in November 2017.

In reality, the project had its origins in my move from Cambridge to Paris in 2005. I had left campus expecting to return. But in 2007, when I opened Le Laboratoire in Paris, I began to see the move as irreversible, and decided to close down my academic lab. For years, my tie to the university was limited to my course, to the twenty or so students with whom I passed a semester back in the United States each spring. I participated on university committees, but I mostly kept my nose down. I had too much to do and to learn. I was curious to understand not just how the things we did and created mattered, but how the *way* we did it—the aesthetic third way—mattered. Eventually this led to the first World Frontiers Forum. I came to Nitin Nohria's office to get his view on where the Forum might bring value to campus, if anywhere. Nohria started our conversation by discussing his day and his work as Dean of the Harvard Business School. Before long, he shifted to the topic of the Forum, and specifically to what had touched him, from talks he had not expected to hear to surprise encounters and works of art. He seemed to link being a contemporary business school dean with experiences that were not typical in the course of a contemporary dean's day.

I had hoped to discuss the future of the World Frontiers Forum, while Nohria clearly wanted to talk about the future of learning.

In our Forum exhibition, videos by the New York–based artist David Michalek captured extremely slow human figure motions of people ranging from infants to an eighty-five-year-old liquor store owner. Five-second movements played in succession over ten-minute film sequences on six three-meter-tall screens. Bells rang softly in a composition by the British composer Brian Eno, who'd created the chime music for the Clock of the Long Now. Nohria asked, "Is this clock actually being made? Did I understand this correctly?" I said it was, and we talked about the Clock for a minute or two. Nohria finished his preface by saying, "We're not missing arguments today; we're missing experience. Our students should be creating such fresh experience and eventually following it to creative careers."

The rest of my time with Nohria was spent brainstorming a campus experiment. Nohria challenged me to create a new experience for the next Forum. In a sense, he wanted to see the creative process I developed in my class applied at the scale of the university. Our new campus experience should involve a few faculty members and several students and lead to some sort of new created thing, which we would exhibit or discuss the following autumn. I might involve whoever made sense at the university, in the arts, in the design school, wherever. If I could give Nohria a proposal, he would manage to raise money for it. "And if you can do this once," he added near the end of our conversation, "you can do it ten times. I'll raise the money for that too."

From our conversation came a collaborative project involving many on campus around the theme of touch.

The consequence of human touch on nature—as on practically everything on the surface of the planet—leaves an imprint that increasingly alters the quality of life. *Touch* focused on the disruption to marine life caused by humans discovering, exploring, and seeking resources from the vast unexplored reaches of our oceans. The project invited engineering professor Rob Wood, a specialist in robotics and renowned for his pioneering design of robotic insects, to work with students to develop novel soft robotic tools that David Gruber, a marine biologist, would bring to practice in ocean exploration. The project had started a couple years before at a National Academies weekend retreat, where scientists, artists, designers, engineers, and others had gathered to think up innovative projects around the theme of the Deep Blue. I'd invited Doug Aitken to join us at the retreat. With Bridgitt Evans, president of VIA Art Fund and cofounder of the Frontier Art Prize, I had started to talk to Aitken about his candidacy for the inaugural award. Now, as recipient of the 2017 Prize, Aitken was in the process of creating an installation made of Maldivian sand—decomposed coral—on an island in the southern Maldives that was fast disappearing with the rise of ocean water.

From the installation, made by Maldivian children and workers, visitors would climb down a stairwell and swim to underwater pavilions that Aitken had prototyped in his first brief exhibition off Catalina Island a year before. The Maldives project was a frontier work of art at a frontier of human civilization—and now with the work of Wood and Gruber, a site where people touched frontier science as well.

We proposed to exhibit *Touch* at the 2018 World Frontiers Forum by having Wood and Gruber speak of the project over an ocean life sustainability dinner.

The dinner came from the mind of a Harvard student, Nicholas Digiovanni, who had, earlier that fall, taken one of the most popular maker classes on campus. The Science of Cooking class is taught by School of Engineering and Applied Sciences faculty David Weitz, Michael Brenner, and Pia Sorensen. It brings top chefs from around the world to campus, where they inspire students to go into labs and learn how to make various culinary wonders while learning the science behind them. Through the course, Digiovanni had been introduced to the cuisine of Dan Barber, chef of Blue Hill restaurant in Manhattan and at Stone Barns. Eventually, Digiovanni approached me with the idea of hosting a dinner inspired by Barber's vision at Café ArtScience around a sustainability theme. One of Digiovanni's interests was food scraps. After talking it over, we decided to integrate his dinner idea into the principal World Frontiers Forum aesthetic experience of *Touch*.

Chef Allen Campbell, once the personal chef of the athlete Tom Brady and the Brazilian model Gisele Bündchen, imagined with Tenzin Samdo and our entire Café ArtScience team a meal around the theme of ocean food scraps. Chef Enrique Olvera of Mexico City's famous Pujol, which was ranked the twentieth best restaurant in the world at the time, joined the creative experiment as the food pioneer of the 2018 World Frontiers Forum, alongside Juan José Ignacio Gómez Camacho, the Mexican Ambassador to the United Nations and an advocate of ocean sustainability inside the United Nations.

Scraps of plastic and fish waste in the oceans reflect the slaughter of ocean life that has gone on since the late nineteenth century. Ocean "bycatch" caused by dragging fishnets was estimated in 2005 by the World Wildlife Fund to account for the deaths of over one thousand whales, porpoises, and dolphins every day. To shed light on the issue and learn from it, we asked Andrew Witt, assistant professor at Harvard's Graduate School of Design, to get involved. I had met Witt several years before when he was working in the studio of Frank Gehry on the Louis Vuitton museum in Paris. Since then, he had developed a protocol for scanning and characterizing material straps and mathematically designing structures composed of scrap material optimally fitted together into contiguous surfaces. I asked my students to create a culinary work inspired by the work of Witt. The result—developed by way of a transient culture lab experience guided by Ken Ledeen and Jacques Vesery and modeled on the collaborative experience of Wood Day—turned out to be a trapezoidal wooden table. The table, finally made in close collaboration with Vesery, had a surface area that was twenty-six times the size of the small white plate that sat alone at the narrow end (for every shrimp that makes it onto our plates twenty-six are needlessly removed from the ocean). The 26:1 table became part of the ocean life sustainability dinner at the World Frontiers Forum.

We also asked Elizabeth Hinton—author of *From the War on Poverty to the War on Crime: The Making of Mass Incarceration in America*—to join the project. Hinton had been trying to give incarcerated prisoners electronic access to one of her classes on campus. She faced the inevitable obstacles to doing anything brand new on a traditional campus, let alone registering students from prison cells. We thought *Touch* might help her while opening a window into the endless possibilities of such projects on a liberal arts campus. Hinton proposed to make a short film documenting the complex issues of "touch" in a community experiencing high levels of gun violence and to share the film in her talk at the Forum.

The entire project became real and entered an electric conversation on campus as we began to wonder how these aesthetic creations and grassroots creator processes belonged on campus in an institutional boundary-crossing experiment around the viability of turning transient culture labs of the university into aspirational culture labs.

Art, Learning, and Change

Had this kind of learning happened back when I was in college, I might have been a more engaged student. The point of school (primary, secondary, university) was generally beyond me. I felt little motivation to learn what my teachers already knew and finally found it hard to turn away from the life that passed by outside my classroom windows.

While I was not the best student, I was lucky to have had a transient culture lab in my house, as well as a live-in mentor: my dad.

He died while I was writing this book. In his death, I remembered many things I might have otherwise overlooked. Like many creators, I am more prone to paying attention to the present moment and to dreams of tomorrow. Memories of my past remain mostly buried in the wildness of my intuition and beliefs. With my dad's death, these memories grew stronger than they have ever been.

Dad was extremely dignified about dying. The process began around the time I flew back from Paris in the winter of 2012. I was with him when he went into septic shock. We thought then it was the end. My sisters and I hoped my mother might have some good years of travel and of living more freely than she'd ever had—my dad, wonderful father that he was, was not an easy guy to live with. But my mom didn't want him to die and seemed almost to will him back to life, which we all witnessed in her fragility and unwillingness to imagine the future without him. My dad saw this. It was like his final act of love for my mother to hang in there for another five years.

And then he fell walking into a restaurant, and his organs collapsed.

I frequently visited him over the last weeks of his life.

I was struck in a very physical way, as if I were the one on my back, losing weight every day, about to leave the world.

This experience, and others I shared with my dad, led to the words I spoke at my dad's funeral. My sons sat in the front row. I saw in their facial expressions that what I said about my dad gave them a new experience of their own dad. I felt something new arise between my sons and me. I couldn't put it into words. Later that day, at the airport in Jacksonville as I walked with my family to the departure gate, my youngest son said to me, casually, as if he thought we might both be thinking the same thing: "Papa, I don't know if I'll have the strength to speak about you like that when you die."

Thierry spoke from his own heart, exploring, I suppose, my view of his interpretation of the day and maybe wondering whether I might have felt similar emotions so many years before when my dad had lost his own dad. His words had a profound effect on me. They helped me explore my own complex feelings and showed me that a generation had passed, and that I had a different place on earth than I'd had before.

We grow closer by sharing aesthetic experiences, and in a way, we shape a future together. It is this kind of raw, surprising, aesthetic experience we all need to share today.

When it comes to the future, we are, after all, in this together. Thirty years from now, we will all eat, get health care, communicate, travel, learn, and work in radically different ways. We will do these things sustainably or we won't do them.

Graceful change is going to take unprecedented collaboration. We will not figure out a new sustainable way of eating, for instance, without benefiting from discoveries at the frontiers of biology, energy, transportation, health care, design, engineering, and more.

Beyond this, what is most true about this collective future is that we all have to want it. We will all have to change how we live on the planet, not just a few of us. The future is not so much a matter of invention, or even of pioneering discovery, but of collective desire, from the most to the least fortunate.

This will require a cultural conversation. We need to get to know each other: to say things that express what we think and feel, not what we think others want us to say; to listen to what others around us are saying; to intuitively hear each other without judgment and change as a consequence. We have to care what others think because we're not alone.

This engaged, expressive conversation is not about selling products. It isn't about entertainment or politics. It's what the best scientists do at the frontiers of science and the best artists do at the frontiers of art.

Cady Coleman is a retired NASA astronaut and United States Air Force officer. She has traveled on the space shuttle three times (in 1995, 1999, and 2011), the last voyage being an extended mission of 159 days. On her trips, Coleman did pioneering science experiments and personally led the deployment in space of the Chandra X-ray Observatory (one of the so-called Great Observatories, along with the Hubble Telescope). Coleman spoke of this experience near the end of the first World Frontiers

Forum, while sharing images and films from her time in space. Coleman argued that space travel was critical for humanity in its search for resources and new living environments, while living in space was itself phenomenal. Even getting from one place to the next was an adventure! When you moved through the space station, you couldn't know which way was "up," indeed you had no reason to refer to "up" or "down" other than to need a convention so that you met people in the corridor without staring at them upside down. She and her team invented games. They flew through the station in choreographed ways just to see what it looked and felt like. They slept, ate, and relaxed as they never had. Coleman brought her flute into space and played it there. Once, she played it live while the rock musician Ian Anderson of the band Jethro Tull played a flute along with her from Earth. She also played in a band with other astronauts, including Chris Hadfield, the Canadian astronaut who once sang and played a rendition of David Bowie's "Space Oddity" on his guitar in space. Music-making with her astronaut friends helped Coleman share her experience as a pioneer among pioneers and relive it freshly. She told us you never looked at the planet the same way after being so far from human civilization that you could point to Earth as a single finite object outside the window. Coleman ended her talk by describing the experience of driving back through the forest of western Massachusetts on her way home following her last mission in space and reuniting with her husband, Josh Simpson, a talented glass artist whose work she had carried into space, as if her life had started all over again.

Coleman's exploratory life captures the spirit of creating what others desire and nobody has created before. Mindful, intuitive, expressive—or aesthetic—experience characterizes the lives of our greatest artists and pioneers of all kinds. It also characterizes every able mind whose future is at stake.

By the aesthetic path, humanity has survived, continually reinventing its future in ways that benefited more than just a few. It has always been in our interest, whenever we found ourselves at an edge, to pay attention to more than just ourselves.

We have every reason and opportunity to follow this same path now in creating a future we collectively want.

Illustration Credits

Page 46	Figure 1:	© Marc Domage
Page 47	Figure 2:	© Veronique Huyghe
Page 48	Figure 3:	© Phase One Photography
Page 49	Figure 4:	© Phase One Photography
Page 168	Figure 8:	© Steven Keating
Page 169	Figure 9:	© Neri Oxman
Page 170	Figure 10:	© Yoram Reshef
Page 171	Figure 11:	© Yoram Reshef
Page 172	Figure 12:	© Michel Figuet
Page 246	Figure 16:	© Florent Déchard
Page 247	Figure 17:	© Phase One Photography
Page 248	Figure 18:	© Phase One Photography
Page 249	Figure 19:	© Phase One Photography
Page 250	Figure 20:	© DR—Incredible Foods
Page 251	Figure 21:	© Phase One Photography
Page 252	Figure 22:	© DR

Figures 5, 6, 7, 13, 14, and 15 on pages 92, 134, 167, 177, 184, and 204 by Bob Roman / PaneVerde Design and Technology

Acknowledgments

Many made this book possible. The project began with an editorial I wrote in October 2014 for *Wired* Magazine, "American Schools Are Training Kids for a World That Doesn't Exist." Arthur Cohen, a visionary of cultural management, gave me powerful advice in writing the editorial, helping it catch the attention of Jeff Silberman, a literary agent based in Los Angeles. Jeff and I met a year later over lunch in Beverly Hills, and the book started to take shape, first as a proposal, aided by the advice and comments of Tom Skalak, a leader of the Paul Allen Foundation in Seattle, and ending up with the brilliant editor Gillian Blake at Henry Holt. Thanks to her guidance and the thoughtful reads of assistant editor and novelist Ryan Smernoff, the book came to be. As it did, many others pitched in. Teresa Amabile, a colleague at Harvard University, read sections of the evolving manuscript and, with her husband and coauthor Steven

Kramer, pushed me in clarifying directions. The artist Doug Aitken put me in touch with the UCLA neuroscientist and creativity expert Robert Bilder, with whom I had a helpful conversation in LA. Eventually, Bob offered critical feedback on all things neuroscience related in the book—as did the talented neuroscientist Andre Fenton, who visited with me while filming a series of documentaries for Nova in Boston. Most of the pioneering creators whose names appear in these pages contributed to my portrayal of their thoughts and lives. A book like this is grounded in an experimental life, and in mine many have been generous beyond belief, including Bridgitt Evans, Cindy and John Reed, and Bob Carson. Without their support I would not have written a single page. A special thanks to each of them, and to my family for their unending patience and support.

About the Author

David Edwards is a creator, writer, and educator. He teaches at Harvard University and is the founder of Le Laboratoire in Paris, France, and Cambridge, Massachusetts. His work, which spans the arts and sciences, has been featured prominently in the international media and is at the core of the international artscience movement. He lives with his wife and their three sons in Boston.